The Tone of Our Times

Leonardo

Roger F. Malina, Executive Editor
Sean Cubitt, Editor-in-Chief

New Media Poetics: Contexts, Technotexts, and Theories, edited by Adalaide Morris and Thomas Swiss, 2006

Aesthetic Computing, edited by Paul A. Fishwick, 2006

Digital Performance: A History of New Media in Theater, Dance, Performance Art, and Installation, Steve Dixon, 2006

MediaArtHistories, edited by Oliver Grau, 2006

From Technological to Virtual Art, Frank Popper, 2007

META/DATA: A Digital Poetics, Mark Amerika, 2007

Signs of Life: Bio Art and Beyond, Eduardo Kac, 2007

The Hidden Sense: Synesthesia in Art and Science, Cretien van Campen, 2007

Closer: Performance, Technologies, Phenomenology, Susan Kozel, 2007

Video: The Reflexive Medium, Yvonne Spielmann, 2007

Software Studies: A Lexicon, Matthew Fuller, 2008

Tactical Biopolitics: Art, Activism, and Technoscience, edited by Beatriz da Costa and Kavita Philip, 2008

White Heat Cold Logic: British Computer Art 1960–1980, edited by Paul Brown, Charlie Gere, Nicholas Lambert, and Catherine Mason, 2008

Rethinking Curating: Art after New Media, Beryl Graham and Sarah Cook, 2010

Green Light: Toward an Art of Evolution, George Gessert, 2010

Enfoldment and Infinity: An Islamic Genealogy of New Media Art, Laura U. Marks, 2010

Synthetics: Aspects of Art and Technology in Australia, 1956–1975, Stephen Jones, 2011

Hybrid Culture: Japanese Media Arts in Dialogue with the West, Yvonne Spielmann, 2012

Walking and Mapping: Artists as Cartographers, Karen O'Rourke, 2013

The Fourth Dimension and Non-Euclidean Geometry in Modern Art, revised edition, Linda Dalrymple Henderson, 2012

Illusions in Motion: Media Archaeology of the Moving Panorama and Related Spectacles, Erkki Huhtamo, 2012

Relive: Media Art Histories, edited by Sean Cubitt and Paul Thomas, 2013

Re-collection: Art, New Media, and Social Memory, Richard Rinehart and Jon Ippolito, 2013

Biopolitical Screens: Image, Power, and the Neoliberal Brain, Pasi Väliaho, 2014

The Practice of Light: A Genealogy of Visual Technologies from Prints to Pixels, Sean Cubitt, 2014

The Tone of Our Times: Sound, Sense, Economy, and Ecology, Frances Dyson, 2014

See http://mitpress.mit.edu for a complete list of titles in this series.

The Tone of Our Times

Sound, Sense, Economy, and Ecology

Frances Dyson

The MIT Press
Cambridge, Massachusetts
London, England

MIT Press books may be purchased at special quantity discounts for business or sales promotional use. For information, please email special_sales@mitpress.mit.edu.

This book was set in Stone Sans and Stone Serif by the MIT Press. Printed and bound in the United States of America.

Library of Congress Cataloging-in-Publication Data is available.
ISBN: 978-0-262-02808-0

10 9 8 7 6 5 4 3 2 1

To my family:
large and small, here and there

Contents

Series Foreword

Leonardo/International Society for the Arts, Sciences, and Technology (ISAST)

Leonardo, the International Society for the Arts, Sciences, and Technology, and the affiliated French organization Association Leonardo have some very simple goals:

1. To advocate, document, and make known the work of artists, researchers, and scholars developing the new ways that the contemporary arts interact with science, technology, and society.
2. To create a forum and meeting places where artists, scientists, and engineers can meet, exchange ideas, and, where appropriate, collaborate.
3. To contribute, through the interaction of the arts and sciences, to the creation of the new culture that will be needed to transition to a sustainable planetary society.

When the journal *Leonardo* was started some forty-five years ago, these creative disciplines existed in segregated institutional and social networks, a situation dramatized at that time by the "Two Cultures" debates initiated by C. P. Snow. Today we live in a different time of cross-disciplinary ferment, collaboration, and intellectual confrontation enabled by new hybrid organizations, new funding sponsors, and the shared tools of computers and the Internet. Above all, new generations of artist-researchers and researcher-artists are now at work individually and in collaborative teams bridging the art, science, and technology disciplines. For some of the hard problems in our society, we have no choice but to find new ways to couple the arts and sciences. Perhaps in our lifetime we will see the emergence of "new Leonardos," hybrid creative individuals or teams that will not only develop a meaningful art for our times but also drive new agendas in science and stimulate technological innovation that addresses today's human needs.

For more information on the activities of the Leonardo organizations and networks, please visit our websites at http://www.leonardo.info/ and http://www.olats.org.

Roger F. Malina
Executive Editor, Leonardo Publications

Acknowledgments

I would like to acknowledge my editors at the MIT Press, Douglas Sery and Sean Cubitt, for their very generous advice and support over the course of this project. For their invaluable contributions to this book, many thanks to Amanda Stewart and Helen Grace, and for conversations and encouragement along the way, special thanks go to Douglas Kahn, Bronwyn Holland, and Adam Lucas. For the opportunity to present my research, thanks go to the National Institute for Experimental Arts, Sydney, Australia; Jason Stanyek; Michael Gallpoe; and the Department of Music at New York University. For financial assistance to begin this project, thanks go to the University of California, Davis. Much of the inspiration for this book has come from the sound community in Sydney, and in this context I would like to thank Jim Denley for many informative conversations, and organizations such as Now Now, the People's Republic, Serial Space, Splinter Orchestra, and the New Music Network, which fund and regularly host experimental music and sound events. Finally, thanks to family and friends for all their support.

Introduction

> Isn't the philosopher someone who always hears but who cannot listen, or who, more precisely, neutralizes listening within himself, so that he can philosophize? Not, however, without finding himself immediately given over to the slight, keen indecision that grates, rings out, or shouts, between "listening" and "understanding": between two kinds of hearing, between two paces of the same [the same sense, but what sense precisely? That's another question], between a tension and a balance, or else if you prefer, between a sense [that one listens to] and a truth [that one understands], although the one cannot, in the long run, do without the other?
>
> —Jean-Luc Nancy (2007, 2)

Cents and sense, eco and echo: these homonyms offer sonorous coincidences that nonetheless indicate a common denomination, a "golden rule," that, when elaborated, would show a more than coincidental relationship between things that we know as "money" and all of the meanings of "sense,"[1] between the two popular meanings of "eco"—ecology and economy; and between space, resonance[2] (as it's generally understood), and sound. The fact that their integration seems coincidental shows just how deeply the connection between space, sense, and eco (meaning the management of a home and "ecology")[3] has been smothered under centuries of de-coupling, abstracting, separating what are essential relations between the extensiveness of space, the perceiving, understanding and experiencing of sense, the resonance of echo, and the flows, currents, and currencies of cents. That these relations have been relegated to the obscure and inaccessible categories of sound and happenstance also demonstrates a form of silencing, a desire not to speak, or to make inaccessible to speech what are in fact the fundamental conditions of human existence. Buried in "sound," made quixotic by a relationship that is too complicated to explain without engaging a poetics tinged with conspiracy, the possibility of bringing economy, ecology, resonance, and space together in the context of habitation seems, in this day and age, almost archeological. Yet the house in which we live, the physical locality in which we reside, and the environment that makes up our habitat (*oikos*) is governed by a process of "making sense" (passing laws, dispensing resources, exacting payments, regulating use) that is administered under the aegis of democracy, but manifests, especially in the current global financial crises, by way of the purely monetary "economy" of currency. It seems almost banal to mention the ties between political speech, media ownership, and financial institutions. It goes without saying that money talks, that those with the loudest voices will have the greatest political influence, and that campaign financing is almost exclusively channeled into broadcasting the voices of select politicians and then analyzing or "making sense" of their speech. It goes without saying not just because it is a fact of life, but also because the fundamental relationship between politics and finance is a fact rather forgotten. Not saying this, it is also possible to deny the relationship between making sense of speech and hearing that speech. Despite the fact that most of the world's sense-making occurs through various technological devices and sounds within physical spaces, the relationship between the output device and the room in which it is heard in the making of sense is rarely questioned. In other words, the actual "sound" of media is ignored, as are the conditions of hearing it. Listening has been tethered to reception, the "demos" of democracy, the open space of public debate (the common) is now wireless; the resonance of voice, wind, material objects, bodies, vegetation, etc. has been exiled to places without people, while discourse and debate now travel the currents of social media, confined to brief snippets of text and thirty-second sound bites, while the populace, ears plugged, dwells within the confines of soundproofed and acoustically regulated walls. Resonance—with its attributes of sympathy, empathy, and common understanding—is reduced to echo: the shallow repetition of the loudest voice. In this day and age, the loudest voice does not necessarily represent the common people, it does not resonate with their wishes, nor engage with their demands, but responds to the markets, to currency trading, flows of money, bond rates, and credit ratings.

Throughout this book I will draw on various scenes, historical moments, artworks, and artistic/theoretical practices in order to situate the reverberative atmosphere that surrounds, encloses, shapes, and sustains us. Each describes a reduction or denial of materiality and the establishment of instruments or operations for the control of space and sense, economy, and ecology, through the sounding of sound. From the moment when Pythagoras, hearing the consonant sounds of hammersmiths, went down into a forge, and upon his exit developed the basis of Western harmony; to Cage's experience in the anechoic chamber; to the relocation of the stock market from street market to building (and now to computer screen); and to the phenomenon of the "people's microphone" adopted by Occupy Wall Street protesters in Zuccotti Park, we find policies and practices of exclusion: the sound of one hammer ignored, and with it, the materiality of noise; the sounds of the world muted, so that an immortal and transcendent interior humming could be established; the rabble of the street that was once the "stock market" enclosed, and with it the transparency of financial trading; amplification at mass rallies prohibited, and with it another pillar of the debt economy stabilized. As allegories, these scenes and moments offer insight into the present collapse of systems, both economic and ecological, by suggesting not just a tale of origins but also a set of practices that shape contemporary life. The sounds of the forge and the rabble of the markets have been muted, rearticulated, and transformed through the monotones of media, "volume" trading on Wall Street, and the gibberish of political speech. The flows of sound may have been reduced to a trickle; however, those of power have only increased.

Chapter 1 journeys into the forge where Pythagoras "discovered" the monochord and in the process set Western harmony on the road to abstraction. The monochord outlined what were "harmonic" and therefore "correct" acoustic intervals, according to mathematical, and therefore rational, relations. This simple equivalence between acoustic and mathematical relationality was to spawn a cosmology that defined the tone as a relation rather than a sound, the musical note as a discrete unit of sound, and harmony itself as a manifestation of divine order. There were, however, a multitude of problems with Pythagoreanism—not least of which was the debate between unity and multiplicity, and music's membership in either category—issues that would occupy music scholars for centuries. The question of music's status as a multitude or a magnitude, and indeed the ontological status of multiplicity as opposed to unicity or "the One," would prepare the ground for the increasing "mathematization" of the universe. It would also designate an area of being as a "remainder": the incommensurable, and incommensurability per se, that resulted from the profusion of ratios that flourished as Pythagoreans attempted to reconcile dissonance in music within a rationally ordered universe. The process of reconciliation, I argue, especially when coupled with the desire to resolve the ontological difficulties of Christian Trinitarianism, would produce the metaphysical, theological, and musical coordinates of paradox—not as a particular contradiction, but most importantly as a mode of thought.

In chapter 2, I read Giorgio Agamben's *The Kingdom and the Glory* and Daniel Heller-Roazen's *The Fifth Hammer* together to elucidate a correspondence between the development of modern music, Christian Trinitarianism, and governance. Specifically, I trace the negotiations involved in reconciling the apparent paradox of the three in one, of a multitude in a unity, the independence between God's will and action, and finally of the relationship between human piety and free will. The "harmony" that co-joins these categories is both social and musical, sonorous (filled with sound) and powerful (effecting organization and subjugation.) I argue that the accommodation of irrationality and incommensurability at the highest level of imagination (God, the infinite, cosmic time, and space) produces a form of "cognitive dissonance." This enables governance as an institution and governments in democratic nations to sustain massive contradictions without falling, like a house of cards. Thus the massive inequities that are at the moment being debated (for instance, the salaries of CEOs) or, the unimaginable debt that countries have accumulated; or, the unknowable consequences of climate change; or, the possibility of sovereign default, unprecedented depressions, and perhaps even the end of capitalism—these unthinkable possibilities are already set within metaphysical systems that are accomplished at performing paradox. To install such a logical glitch at the very beginnings of Western thought was a major accomplishment—we see this retrospectively of course—because it enabled the progression through the centuries of something like a metaphysical wormhole into which the materiality of power, the force of legislation, the division of populations and their enslavement—first to monarchs and now to institutions of democratic governance—is, if not unsayable, then virtually unassailable. Timothy Morton has written that it is easier to imagine the end of the world than the end of capitalism (Morton 2010, 101). I should add that this includes the end of democratic governments, without the structure of which we are faced with what has been designated as "the void" of anarchy.

In this light, the uproar caused by so-called jagged blips, or "ear-piercing noise" of composers like Stravinsky in the early twentieth century, was not necessarily an uproar over dissonance. The affront ran far deeper than musical sensibility, all the way back through the centuries to those initial determinations regarding not just the ontological foundations of Western thinking, nor the theological doctrines of Judeo-Christianity (ontotheology), but the sleight of hand that introduced a conceptual vacuum at the heart of thinking and especially at the heart of the thinking of being. To elaborate, I discuss the concept of tone via Alain Badiou's concept of the "count-as-one," which is relevant to the formation of the concept of tone as a discrete but nonmaterial entity.

In the development of Pythagorean harmony, the tone was removed from its material sounding as consonant or discordant, and considered, like number, as a pure relation. This allowed it to pass between actual materiality—avoiding any designation of pitch or frequency, for instance, and operating instead on a purely arithmetical plane—as a relationship between different pitches or frequencies that would be considered harmonic or discordant. Thus the concept of the tone has always enjoyed both an abstraction and a certain materiality. Its differentiation from a note, its oscillation between the materiality of sound and the structure of music, can be seen in the multiple meanings that "tone" suggests. We think for instance of vocal tone, the tone of the situation, atmosphere (as in "atmos"), etc., as something beyond specification, both a sound and a meaning, that cannot be translated into language, but yet it is fully understandable. The nuanced nature ascribed to tone arises, one could almost say directly, from this ontological tug of war between the individual unit, the One, and the multiple.[4] Within contemporary musical practice and sound theory (including acoustics as well as scientific and computational methods of sound analysis), the same problematic operates between materiality (the frequency and timbral aspects of a certain sound, for instance) and meaning in an extramusical sense. Without the structure of music, without the enclosure of tones within a specific system of notation, the musical sound *qua* sound, presents its materiality, its multiplicity, and therefore its anteriority to music *qua* music.[5]

The relational character of the tone enabled it to act as an agent—an operator—in the creation of harmony. Having been incorporated within Pythagoras's harmony of the spheres, the tone also shared the "mystery" of the theological doctrine of the Trinity, as music itself was now a result of the "mysterious" workings or arrangement (economy) of notes, rather than the musical qualities of sounds themselves. The question of whether music was a multiple (a collection of notes for instance) or a magnitude, was resolved in favor of the latter—that is, a singular identity, indivisible and unique, if only analogically. While Pythagoreanism had its sonorous manifestation in the monochord, Trinitarianism was manifest through the "song of praise," which, as discussed toward the end of this chapter, held within its repetition of divine names (such as "Lord of Lords" or "Holy, Holy, Holy") the echo that would constitute the endless praise of divine worship. Echoing into the spheres, harmonizing the movement of the planets, creating a symphony for divine audition, and affirming the infinite time, space, and power of divine being, the song of praise can be seen as a sanctified instance of poetic speech, that, unlike its secular counterpart, lays claim to cosmic space through sound. The liturgical function of worship or acclamation sets this form of appropriation-through-sound within human reach. Combined with the "administrative" operations of the Trinity—whereby, as Agamben theorizes, power and the administration of power are separated through the division between the Father and the Son—the song of praise and liturgical worship in general create a powerful instrument for not just the glorification of the church, but also the eventual glorification of the governmental apparatus itself.

This process is as relevant today as it was then. In the same way that governance has been installed historically through the amalgamation of theological and political power, governance is currently installed through the amalgamation of political and financial power: the Sanctus has been replaced by the endless echoes of mediated *doxa*, while the mystery of the Trinity can be found in arcane financial processes that trade in impossible equations that far surpass the Trinitarian three-in-one. In the secularization of the West, it could be argued that the concept of divinity has undergone a process of "techno-gnosis," whereas the infinite cosmos has been usurped by the ethos of progress. To illustrate the movement between the theological, technological, governmental, and financial spheres, I discuss Ben Rubin and Mark Hansen's 2002 sound/art installation *Listening Post*, the first iteration of which can be seen as a work that transforms Internet chatter into a form of liturgical chant, and the second as a sonic exposition of the machinery underlying that process. The parallel relationship that develops between the notion of musical harmony and cosmic harmony on the one hand, and social harmony and theological governance on the other is profound. Despite their seeming disparity, there is a point of contact, of coincidence, between the two that enables a merger between them. This point is the accident, the accidental, the "Pythagorean comma," the irrational, the incommensurable: the tone that cannot be accommodated within the strict mathematics and ratios of cosmic harmony, but yet must be accommodated in order for the Infinite to function as a concept within the ethos of progress and industrial modernity, and in order for theology to enter—and shape—politics at its fundamental conceptual roots. Without the inclusion of the irrational or incommensurable, the symphony of the spheres would sound the same tones over and over; there could be no movement in music, no change, simply the endless iteration of simple and ultimately boring melody. The solution, spanning centuries of musicological, philosophical, and theological scholarship, has been the creation of a mechanism of separation—providing a distance between the mover and that which is moved, the creator and that which is created. That mechanism is the irrational, the incommensurable, discord, "noise," and is the focus of chapter 3.

Noise: Think of two people standing on fractured columns of concrete slab waiting for the jackhammers to stop so that they can continue their conversation. There is a brief moment of relative silence. The two seize the opportunity, condensing their points, abbreviating their style, and shouting the last few words of their sentences as the jackhammers resume after their three-minute break. The strategy fails, as their speech trails off in a series of coughs and splutters, caused by the additional inhalation of particulate matter that fills the atmosphere as the speakers fill their lungs in readiness to shout. An onlooker is momentarily amused by the scene, but this quickly passes into annoyance at being unable to hear what sounded like an interesting debate. She walks within hearing range and suggests they move somewhere quieter. "That's what we were arguing about," says one, "Whether there is any other place." This is the reality of noise—a sonic phenomenon that can invade territories, reconfigure discourse, drown out speech, and interfere with the transmission of meaning. Noise deafens—the more you hear, the less you are able to hear, the less you can hear, the more noise you need in order to hear. Noise undoes its own hearing and in the process multiplies. It is this kind of noise that I would like to keep churning in the background as the scene I have just described is transferred to the sound studio, where noise will be attenuated, the harsher frequencies filtered, the fuller and more interesting or unusual sounds boosted, and varying amounts of reverb and delay added. Here the noise of the construction site becomes material for an audio artwork, or perhaps an experimental music composition. Its sound may be just as loud, its frequencies just as grating to some, but thankfully it will always be subject to the volume control—whether it be from amps onstage or speakers in the living room. It could also provide background for a feature on what is happening at the construction site—a new development, praised by some, contested by others, the subject of planning laws and regulations, aesthetic precedent, environmental assessments, etc. The processes of filtering industrial noise just described can also act as an allegory for the control and containment of noise per se. I have capitalized the "N" here in order to differentiate between local, everyday acoustic "noise," and "Noise" used conceptually, as an engine of difference and, aesthetically, as a form of resistance to stultifying and exclusive cultural values. To some extent, I am following Giorgio Agamben's treatment of the voice (Agamben 1991), in order to emphasize noise abstracted and aestheticized: the way, for instance [N]oise functions as an *urstuff*, the undifferentiated, unpresentable flux from which all information, all meaning, emanates. The interesting, ironic, and for the purposes of this chapter, salient aspect of Noise is that it functions negatively—as unmeaning or nonmusic—yet while positioned negatively, as a sonic and informational opposite, the negativity of Noise is continuously challenged by the materiality of "noise," and with it, the whole project for which Noise as a metaphor stands.

Traditionally, in both the audio industry and informatics, noise is considered a form of interference that impinges on acoustic space and the transmission of meaningful signals. However, the pursuit of noiselessness has also been compromised by internal contradictions arising from the oppositions it activates: dissonance is opposed to consonance; timbre is opposed to tone; noise is opposed to signal; the material world is opposed to the abstract model; difference is opposed to the same; multiplicity is opposed to the One. Conceptually, both the notion of a pure tone and pure signal rely on the suppression of noise and the enforcement of these oppositions; both represent an ideal of musical composition and communication based on a transcendental subject: an ear without a body cocooned in a world of silence, or a medium of transmission that communicates ideas transparently without any loss of meaning nor any influence from the medium itself. This model denies the noise immanent in any signal—be it musical, linguistic, or electronic. The "true" signal is always an approximation—of electrical current, or acoustic frequency, or linguistic description and characterization. As information theorists and audio engineers note, even if noiselessness were achievable, a certain amount of redundancy is necessary in communication, just as a noise-free audio recording is perceived as lacking "warmth" and consequently less musically satisfying.

In the continuum between noise, audio ambience, and potential music, noise undergoes various forms of "sonification," depending on the kinds of filtering devices it has passed through. Although this is true of noise in its informatic and sonic sense, I will confine myself to sonic "noise" noting as I do the apparent devolutionary tautology that particular phrasing strikes, since it makes clear that the idea of sound no longer contains the subset "noise": rather noise has escaped the boundaries of the sonic or acoustic in ways that directly confront modernist noise aesthetics (or what is often referred to as "musicalized noise") where noise is simply a material for sonic *poiesis*, for bringing into being a latent musicality, as if noise were "raw" and just in need of a little technical or compositional refinement. Its use as a source of both sonic material and musical inspiration is commonplace:

think of the mimetic birdsong in Debussy, industrial sounds in Stravinsky, environmental ambience in Cage, signal noise in "glitch" music—the list goes on. In all cases, sound from the world enters and disturbs the closed system of Western tonality or the false mythology of neutral media. Contemporary "noise music" goes further, rupturing both tonality and eardrums as volume and frequency range are used to vibrate the listeners' organs, rattle their bones, and transform sound or noise into a physical force that challenges the audible as a separate sensory phenomenon. In this context, noise, as a sonic material, is both transgressive and generative, and, importantly, tied to the material, sensual world. However, there is also a tendency to glamorize noise, to run headlong into its unspeakable flux, to become transfixed by its transgressive sheen. Aestheticized noise provides an apology for acoustic bombardment by framing noise within the walls of the gallery or auditorium, which are not only insulated against acoustic noise, but also protected from critique—from naming noise as (annoying) noise. This is, I suggest, part of a process of capitalizing on noise.

What is left then after these conceptual capitalizations? Perhaps no other origin myth reveals the importance of noise and its undercover agency, so to speak, than the big bang. The term itself coined satirically by Fred Hoyle contains many of the paradoxes that are integral to the concept of noise. First of all, the big bang is *both* the sound of an event (the beginning of the universe) *and* the event itself. Second, this noise initially attracted the attention of astronomers as interference, as something they wanted to eliminate in order to proceed with their investigations. This is part and parcel of the paradox of noise: noise is by definition unwanted. Third, the onomatopoeic character of the term itself counters the infinity it would describe. The fricative "bang" rapidly diminishes after its initial vocalization, having no echo, no reverberation in a second syllable, no repetition of the second word as one would find, for instance, in the Sanctus. Again, this is a characteristic of noise and also onomatopoeia, and it is not coincidental that the two are so closely allied in language, as the majority of words for noise are onomatopoeic. This association draws a limit around discourse, for how can we talk about noise, given the repertoire of onomatopoeic terms on hand, and still be taken seriously? At the same time, however, the awkward name for the origin of the universe can be seen as an echo of modernity, since it sets the "bang" apart from all other sound, turning what is usually considered to be an ambience—composed of multiple sounds that are usually only identifiable by the object that creates them—into a singular entity or event. In its singularity, the big bang designates a beginning to the universe and in so doing automatically brings into play the possibility of an end. The Infinite is thus touched by finitude; the infinity of theism surpassed by the doctrine of perpetual growth. That this growth is periodically ruptured by market "corrections," recessions and depressions, did not prevent financiers from also naming the unprecedented unification of markets during the 1980s the "big bang."[6]

As mentioned, noise always brings with it the materiality, the physicality, and the actuality of the situation. Noise interferes with processes of abstraction and idealization, with the desire to disregard one part of philosophy (moral philosophy) in order to secure another (rationalism). In this sense, the scene of the forge is a fabrication, since in disregarding the fifth hammer and the irrational ratio it physically sounded (known later as the Pythagorean comma), Pythagoras disregarded and so marginalized and eliminated the hand that held the hammer, the space in which it sounded, and the conditions of sound's production. "Hammer" is both a verb and a noun, and is found in toolboxes and musical instruments alike. Its most basic meaning as a verb is to beat or drive, with the connotation of laborious attempts to complete a work, a plan, or an argument. In this sense, jettisoning the fifth hammer perfectly symbolizes the persistent drive of ontotheology and scientism to eliminate the pause, discontinuity, or the incommensurable that interrupts rationality. There are two sides to this noisy coin, though, for the irrational inevitably insert a degree of unpredictability into the mix, opening the way for external forces (extreme weather, rogue trading, mass uprisings) to wreak havoc on the very institutions that have denied their existence. At the same time, however, the accidental provides exit clauses for the failure of human and divine creation: natural destruction, cataclysm, disasters that could never be predicted, but also wars, depressions, environmental disasters, industrial accidents, and so on. Noise allows these episodes and events to escape judgment at every level of governance. In the context of these evasions and recuperations, I discuss Edwin van der Heide's *LSP* (laser sound projection, 2012),[7] and Ryoji Ikeda's *Datamatics* (2006–2008), which literally work through noise as both signal and musical material, as a sonic representation of cosmic immersion and the pure, rational tones of the monochord. In doing so, the artists illustrate the dynamic process of incorporation and resistance that the aesthetics of noise exemplifies.[8]

In chapter 4, I explore a similar process to those used in *Listening Post* through the field of affective computing, but instead of "voicing" online communication through art, I look at the computation of tone—in this context, the tone of the voice—as a means of gauging a speaker's emotional state in human computer interactions. Both renderings of the voice and its tone—the computational and the aesthetic—impinge upon and renegotiate one of the central questions of posthumanism—one that we see constantly worked over in science, art, and culture, namely whether and how there could ever be machine subjectivity. In the quest to enhance remote communication, for instance, the computation of vocal tone is integral, yet tone itself escapes easy definition: is it best understood as pitch, frequency, or vibration? Does tone act as an emotional trigger because of its sound, its musicality, or its meaning? Tone is at once a musical, mathematical, sensorial, philosophical, and technological arbiter; yet at the same time, it remains an enigma across all fields. In its oscillation between radically different discourses, tone introduces a complexity into media ontologies that are based on the unit—be it code, signal, or subject. This is undoubtedly nowhere more obvious than in the attempt to quantify, compute, and reproduce affect in the tone of the voice. True, most of us understand each other's tone; we have internalized what a gruff or dismissive or joyous tone is; we know how to read the various modulations and vocal pitches that reveal the current emotional state of the person we're speaking to. However, while tone might carry an enormous amount of meaning, trying to characterize, let alone quantify, just what tone is is extremely difficult. The extent of this difficulty, the depth of the problems, assumptions, and worldviews it poses lies in the unique intersection of sonic, linguistic, and musical elements that cohere in vocal tone.

Affective computing identifies and measures a person's emotional state by tracking bodily indicators such as heart rate, galvanic skin response (sweat), breathing rate, gait, facial expressions, and the like. The tone of the voice is considered to be one such indicator, and its computation, by research groups such as MIT's Affective Computing Group, is indeed a Noah's Ark of the voice, a massive encyclopedic venture reminiscent of earlier linguistic ethnographers who used recording devices for analyzing exotic languages. However, at almost every stage in this process, we see a withdrawal of human conversational exchange. For instance, in building the initial database of various tonal inflections, the group chose not to use human actors because actors tend to sound unnatural. Instead, the raw data, the natural speech, is collected from volunteers who navigate an interface guided by "embodied conversational agents" who prompt the user to speak about an emotional experience. In other words, the simulation of affect (by actors) is replaced by the simulation of conversation.

Returning to the familiar calling card of the telephonic voice, the prelude to the remarkable technological feat of transporting the voice from one locale to another—we find that each new mode of telepresence: from wired telephones to wireless radio and now to wireless media raises the question of presence, hidden in the ubiquitous, unchanging greeting "Hello," followed by the peculiarly telephonic protocol of asking, "Are you there?" This fundamental question has been answered by two related, recursive, operations—the increasing sophistication of communications technologies, on the one hand, and the adaptive techniques developed by speakers, on the other. In this chapter, the question of "Are you there?" interweaves a conversation between Derrida and Stiegler on the one hand, and the flailing attempts at dialogue by REA (short for "Real Estate Agent"), an "embodied conversational agent." Through these conversations, the complexity of everyday questions such as "Can you hear me now?" and "Are you there?" becomes apparent, as does the problematic of (tele)presence.

As I discuss in chapter 1, *Listening Post* uses a synthesized, computerized voice to "humanize" the somewhat alien medium of text-based conversations on the Internet. The transformation of chatter into chant through art and sound is of the same nature as the transformation of vocal tone into useable, quantifiable information. Both are processes for eliminating noise, and both are also mechanisms for producing, or simulating affect, if not by calling forth the pseudo-transcendentalism of a musical or religious sublime, then by mapping the acoustics of tone onto the contours of algorithms that have themselves been charged with the impossible task of defining affect. But bodies and voices engage with these technologies in ways that trouble the clear transmission of both meaning and affect, evading either metaphysical or computational articulation, disappearing, like a mirage in the desert, as they are approached. To the extent that affect can be delivered in the form of physiological signals or acoustic markers, it can also be performed, worn on the sleeve in a caricature of the affective profiles used to sense, and potentially manipulate, the user's emotional state—an emotional state that, it might be added has through monitoring and surveillance become *disaffected*. As I argue, the technologies that filter, tag, analyze, and codify the voice might retain it as data, but do not hear it as sound. If it is no longer the voice that speaks nor the ear that hears, but an intermediary taking the form of a device occupying a database and "listening" through systems that filter an already enigmatic and ineffable tone from an already technologized, instrumentalized speech, one wonders then how affect can be generated? This question intensifies when surveillance—as a technological other—in fact creates the performance, the "untruth" it is designed to elicit. Indeed, the problem of surveillance and its corollary, dissimulation, cascades through the entire technical, perceptual, and epistemological process—"passing on" and perhaps amplifying a form of acquired autism in the next generation of software. The filtering processes that occur in the computation of vocal tone are not isolated to affective computing, but rather can be seen operating across media culture. These, together with the space of sound, the environment in which voice engages in conversation, and the impediments to or enhancement of its transmission, are the subjects of chapter 4.

Sound, tone, music, voice, and noise: these identify five forms of sonority through which the crises of eco—both economic and ecological—can be read. In chapter 5, I sift these sonorous forms through the sieve of resonance, poetic speech, and in particular, the noise works of sound poet Amanda Stewart and what I am calling the general state of anechoica that Michel Serres interrogates in recent writings. Sound's reverberation, and the separation it implies, is pivotal in defining the "echo" of "anechoic," for only by separating the listener from what Serres calls the "hard sonorities of the world," and the sense of listening itself from the other senses can the reduced sonic habitat of contemporary urban life be established. This separation or filtering occurs through a complex process that Serres outlines in *The Five Senses* and from which I draw heavily in order to situate the anechoic within the "space" of sense—the atmos, if you like, of common sense, which I elaborate through sound. Central to Serres's analysis is the containment of noise and its transformation, or "softening" from the "hard" noise of the given—the sensible—to the soft signs of culture. This transformation occurs by way of enclosure, pollution, degradation, and finally extinction—of land and other species—and is, according to Serres, no less than a form of absolute theft. For Serres, noise begins in the body and emerges into language via the sounds of the earth: the gushing of water, the soft cacophony of rustling leaves, or the screaming of gale force winds. Small sounds or high-frequency energies are quieted by big sounds—low, slow frequencies that are attached to, or rather part of, the movement of the elements. These are the sounds that, according to Serres, belong to "the given," the material world, the sensible as that which is "designated by the infinite capacity of sense" (Serres 2008, 118). They pass through the body and are filtered by the skin, and in this way, noise and vibration become meaning, hard sounds become soft, and the world becomes bearable.

In the same way that voice and language belong to sound, which moves and circulates through the body and the social, philosophy is moved and circulated by the tonality of speech. According to Serres, philosophy is tethered to the fluidic and tempestuous flow of sound, and sensation is intricately linked to the production of language, even though language softens and obliterates the possibility for our own recognition of this sensation. By linking philosophy to, for instance, the rustling of the leaves, Serres links knowledge and analysis to a simple phenomenon that in itself presupposes an entire natural world: the movement of air, the openness of skin, and the unproblematic capacity to sense this sound as a multiplicity that encompasses every aspect of being, including the social. Serres urges us to write as close as possible "to the full capacity of the senses which is opened up in this place and given to us by the sensible. … The sessile leaves of the poplars write the sensible, say the sensible, to be read here, and heard too" (Serres 2008, 118). He urges us to write, that is, in a voice made from the union between the body's tissues, the ears, and currents of air. It has been very easy for philosophy to ignore this "voice"— to refer to it as "flux," to deem it unpresentable and unsayable, or to cast it entirely within the social, contained and ordered like the "managed nature" that places a human footprint over all existence. Serres calls for us to retune our senses in a way that Murray Schafer did with his provocative and seminal book *The Tuning of the World* (Schafer 1981). However, unlike Schafer, Serres does not proceed by way of cataloguing or analyzing sound, or differentiating between hi-fi (rural) and lo-fi (city). Serres's discussion is not a critique of modernity, and the loss of the sensible that he refers to cannot be recovered by inventing better technologies (as Schafer reintegrates the listener through the use of headphones). For without any intervention, the softness of language and culture drowns out the hardness of the given, and in so doing, it erases its foundation, the "inaccessible totality of meanings" (Serres 2008, 118).

Serres describes this gradual demarcation and quarantining of the senses along with the control and exploitation of the environment using the analogy of the black box in science, but also the boxes that we live in, that have become our habitat. Isolated in houses and buildings, with cities domed by data clouds, we have become deaf to the sounds of the earth, and as a consequence now lack the ability to sense, and this is pivotal to the sustenance of common sense. All that we hear, all we can listen to, is the racket of our own voices—arguing, legislating, pronouncing, warring, and transferring power. What happens when "the racket" substitutes for the environment? When the solipsistic fibers suspending western metaphysics lead us to believe, as Serres writes, "the world is crisscrossed by nothing more than signals" (Serres 2008, 146). The consequences are still unfolding and increasingly dramatic: over the last decade, the common has been appropriated and destroyed, and the driving force has been the production and circulation not of wind or atmospheres, but of other pockets of air, also known as bubbles, filled with fragile promises that are continually bursting. The high-frequency buzz of the collective favors these anechoic circulations, as a result of which the dimensionality of meaning or tone drops out as a kind of material residue. What is lost in the process is, according to Serres, "quite precisely—our common sense" (ibid., 118).

In chapter 6, I return to the now infamous moment when Cage entered the anechoic chamber. While Serres would have been horrified by Cage's experience—because for him being able to hear the body's inner workings implies the silencing of the world, in particular, the absence of movement and sensation, that would normally quiet the "humming of the cells, the hubbub of the organs" (ibid.,106)—for Cage, it was a life-changing experience. Anecdotally, it appears that when Cage heard for the first time the circulation of his blood and the buzzing of his nervous system, he was probably hearing his tinnitus. Tinnitus, associated etymologically with the jingling of coins, is the perfect metaphor for the kind of misrecognitions that a space of no sensation favors. Thus when Cage announced after his experience that, "Until I die there will be sounds. And they will continue following my death. One need not fear about the future of music" (Cage 1961), he may have also mistaken music's future for ours—specifically the future of our listening as it shrinks to fit various enclosures and gets used to the internal racket of jingling, but soundless, senseless, coins. The sound of coinage, of money, or its fiat, circulating and transacting the globe, surfaces in the concept of "financial noise." In tracing the gradual enclosure of the stock market—from the everyday street to Wall Street—we also find the development of the nonacoustic "noise"—of high-volume, high-frequency trading that is part of the financialization of everyday life. Drawing from Maurizio Lazzarato (2011), David Graeber (2012), Christian Marazzi (2011), and others, I outline some of the processes and consequences of financialism, in particular, the existential demands of the indebted subject, exemplified by Helen Grace's (2009) installation *IPO*, in which Grace quantifies, charts, and values her daily emotional life over a nine-month period, transforming affect into a stock market index. Hardt and Negri identify four features of the new subjectivity produced by the current state of institutionalized democracy:

> The hegemony of finance and the banks has produced *the indebted*. Control over information and communication networks has created the *mediatized*. The security regime and the generalized state of exception have constructed a figure prey to fear and yearning for protection—*the securitized*. And the corruption of democracy has forged a strange, depoliticized figure, *the represented*. (Hardt and Negri 2012, 7)

To these figures we might add "*the eco-guilty*"—produced through the internalization and individualization of the general recognition of anthropogenic environmental degradation, which often enforces and empowers the four figures Hardt and Negri describe. This new subjectivity, an outflow of what we might call the postmodern condition, is experienced on a subjective level as an overwhelming *condition* of debt, which has become necessary for social life. The shift in capitalist production from the factory floor to everyday life, the rise of "I-culture," the culture industries that I mentioned previously, globalization and the 24/7 work week, have absorbed every aspect of creative and productive, not to mention affective, life.[9] Within this sphere, social rights become social obligations—or more correctly, social debts—but without the (civil) "society" that would once have ameliorated the guilt associated with individualized debt obligations. For Lazzarato, the transformation of social rights into debts and beneficiaries into debtors is part of a program of "patrimonial individualism" that requires continuous "work on the self" in order to produce what he calls "indebted man."[10]

Serres writes:

> I do not know if talking of filters will help us understand how thunder, noise, and the vibration of sound waves (whether audible or felt through skin) subtly become meaning. There is no reason to discount the hypothesis. The question of knowledge, of the sensible and of language is located somewhere on the graduated spectrum of this fan, somewhere in the range it encompasses between hardness and softness, this partitioned, compartmentalised distance, strewn with obstacles, twists and turns, and clear pathways. A box within boxes where the sound of cannon-fire gradually transforms itself into a whispered confidence. (Serres 2008, 115)

Over the last decade, cannon fire, as a weapon of mass destruction, has been transformed via media into "weapons of mass destraction,"[11] which now operate as financial weapons of mass impoverishment, locked up in algorithms that no one can understand and, as such, are both senseless and beyond scrutiny. The eerie silence of a stock market crash, of low volumes and high volatility, are a reminder that quarantining the market from the street, installing financial exchange within anechoic determinations, literally produces the kind of ratios that Pythagorean harmony was eventually unable to sustain. Similarly, the "echoing" media are a reminder of the consequences of speech that are endlessly repeated. The myth of Narcissus and Echo is prescient here, because Echoes chattering, her long stories, her ability to distract, becomes her own worst enemy—in repeating the final words of another, she is destined only to chatter. Chatter, related to twitter and gossip, has only recently been associated with electronic communications and threats against national security—the term has a much older association with rumor, babble, and hearsay, and describes meaningless, unfounded, and incredible speech. In the midst of the economic and environmental crises, the racket can be seen as the sonic manifestation of extreme governance by financial means (the financial racket or financial noise) in contradistinction to the raucous rabble of street demonstrations and riots, that have a "voice" in the people's microphone. These form part of a global movement, one that, Hardt and Negri describe as being heard "on the lower frequencies: [that] are open airwaves for all" (Hardt and Negri 2012, 3). Indeed, what occurred in these struggles was, they infer, something unique. A new common sense, one that no longer accepted the threats of financial Armageddon from politicians and chosen economists, that grew from the rejection of representation, the development of democratic participation, and the occupation of a nonmediated common space. The physical act of occupying and reclaiming the common, then, developed a new sense, a new way to *make* sense, and to make sense of the current crises (ibid., 6), and the spacing, pausing, and listening involved in and produced by the people's microphone is an articulation of this common. Its echoing punctures, or inserts a comma, a pause, in the covering up of power. Its silence denotes a silencing, and its repetition is the insistence of echo—that voices will be heard and speech passed on. Similarly, the sound cultures that are developing across the globe, listening in old warehouses to performances that often consist in little more than the very slow modulation of a single tone, these cultures form a sense, and practice of sensing, that it seems to me offer a way to the "sensible," and perhaps a way out of the thinking that is in a constant state of leveraging Eco.

1 Endless Praise and Sweet Dissonance

In demarcating sound and the voice from vision, Nancy uses the term evocation:

> We should say that music [or even sound in general] is not exactly a phenomenon; that is to say, it does not stem from a logic of manifestation. It stems from a different logic, which would have to be called evocation. ... Evocation: a call and, in the call, breath, exhalation, inspiration and expiration. ... What comes first not the idea of "naming," but that of a pressure, an impulsion. (Nancy 2007, 20)

Evocation is a call, a summons, to the gods in Roman times, to the spirits of past existences. The term is associated with an indirect communication, not necessarily a calling to someone or something, but to a being, a deity, or perhaps an evil spirit, that does not have a sonorous voice, but rather, a silent and invisible presence. The "pressure" of which Nancy writes is related to this drawing up from the depths, or calling down from the heavens, as much as it is related to the breath, to the brief exhalation that accompanies a call, and to the impulse behind the summons. In other words, evocation has its exhilarating and spiritual dimensions. Somewhere in every call there is a summons to a nonhuman being, as well as a brief movement of breath from inside the body to the surrounding atmosphere. The two axes of evocation might be experienced when listening to an orchestra and feeling, like a light massage, the eardrums vibrate. "Music to the ears" is not just about hearing a beautiful melody, feeling the fullness of orchestration; it is also about the very fine touch upon the ears that the orchestra and its sounds transmit. In this sense, it is a pulse, a kind of fluttering, that goes beyond even the sonic. This double sense of exhilaration—as both an exhalation of air from the lungs that accompanies any vocalization (but that also connects individuals with the atmosphere they inhabit in the most intimate of ways), and the feeling of being uplifted, has been transformed through the Judeo-Christian tradition, removed from animistic or pagan mythologies, and set within religious doctrines that utilize the immediate relationship between the physical and emotional or cognitive ideation. Here, every breath becomes God's, every breath a summoning of spiritual guidance or favor. But more than this, the evocation not only summons the gods and thereby multiplies or extends the individual, it performs this transformation by inscribing a logic based on mathematics. According to Heller-Roazen, this logic will determine the cosmology of the West, the cosmological path that has led to the present ecological and economic crises (Heller-Roazen 2011). This is the mathematics of harmony and the structure of the tone, which has influenced musical, scientific, philosophical, and religious doctrine from the time of Pythagoras. However, like the hills that sounded Echo's diminishing voice, the Pythagorean monochord—the basis of his theory of harmony—also contained within it an element unrecuperable within the mathematical formula that would, ideally, provide a model for the numerical quantification, transcription, and notation of nature.

Pythagoras "discovered" the monochord quite by accident. According to Heller-Roazen (2011, 12), he passed by a forge where he heard hammersmiths ringing out a consonance as they pounded the metal into objects. At the time, there were five hammers; however, the fifth hammer, because it lacked consonance or because it didn't fit into the numerical formula that Pythagoras was devising, was ignored:

> Although Pythagoras wished not to include the last hammer in his equivalences of noise and number, he nonetheless perceived it. Boethius leaves little doubt: immobile before the forge, "as if spellbound," the sage "overheard the beating of hammers somehow emit a single consonance from differing sounds." ... Its consequences were to be lasting. Pythagoras might well return to his research; he might well reject all instruments. Dimly or distinctly, if only for a moment, he had nonetheless perceived a being without measure. (Heller-Roazen 2011, 17)

It may have been that the brute tool was considered "worldly" and therefore corruptible, whereas the model of the fourfold was a principle of nature and therefore eternal, immutable, and necessary. It could also have been that the discord of the fifth was a testament to its lack of necessity—by a perverse kind of logic, Pythagoras determines that the actual sound is found to be incommensurable with the model, proving its marginality. As an allegory, however, the scene at the forge is instructive: thinking of the sound's environment rather than the actual tone, we find that this scene in which the unaccountable is ignored occurs in an industrial setting, in what would be the precursor to the industrial age. One can imagine a scenario replete with all the elements that operate today in sonic capitalism: a thinker, a member of the privileged class, visits (as in "goes down" to) the forge, where he notices not the men, but the instruments, and not the purpose of the instruments, but rather the sounds they make. It would be a stretch to suggest that Pythagoras was a very early pioneer of musique concrète—although there are some methodological similarities, particularly in the separation of sound from its material source. More to the point, the hearing of sounds, or rather noise, as music, or at least as exhibiting a feature of music (that is, consonance), and the distinction between concord and discord, tonality and atonality, occurs within a social environment characterized by the activity of work. The environment in which Pythagoras makes his discovery is already one predisposed to a certain distinction: between reflection and labor, choice and necessity, privilege and lack thereof. Acoustically, the forge was able to produce such clear consonance no doubt because it was enclosed, walled by acoustically reflective surfaces such as bricks, whereas this kind of resonance may have been impossible in the open-air fields and walks where Pythagoras, one imagines, did his thinking. "Forgery," like "racket" (which we will come across later in the book), is instructive in that it means to build or to make, and also to copy, to counterfeit. It has both the productive sense of creation and generation, and the negative sense of unreality, deceit, and fabrication. Industry and commerce, the production of objects, would be impossible without it; however the value of the product is impossible to determine when it is present. Perhaps in that moment of dissonance, Pythagoras heard vibration, the materiality of the hammers, the sound of the metal, the presence of timbre disturbing the pure tones of an evolving harmonics. The fifth hammer was discarded in order to save the model (tonality) from its materiality (sound) but as a necessary consequence, so too was the excess baggage of social injustice, inequity, and the always questionable rationality of ethics, within what was supposed to be a perfectly harmonized cosmos.

The Monochord

The road to abstraction always begins with simplicity: that is its rationale. For the Pythagoreans, the monochord introduced a unit—the tone—which could then be placed in relation to other units (tones) as musical intervals:

> Suddenly, the seemingly endless diversity of sounds acquired a new simplicity. Acoustical intervals were now expressible as arithmetical relations. The proof lay in the reduction of the sound of the octave to the relation of two to one (2:1); that of the fifth, to the relation of three to two (3:2); that of the fourth, to the relation of four to three (4:3). (Heller-Roazen 2011, 14)

This discovery was not confined to music or acoustics but could be applied to the world at large. *Musica*, often regarded as the highest of the arts, could now be represented; but more than this, the invisible, ephemeral multiplicity that is sound could be unitized. Pitch and duration, the main elements in musical composition, could be considered discrete units rather than continuums, intervals, like numbers, made up of "ones" known as tones. While the monochord is often seen as just one manifestation of the broader desire to make the universe literally accountable by ignoring noise, the Pythagorean doctrine, as Heller-Roazen shows, resists the temptation to obliterate materiality altogether by grounding mathematics in experience and insisting on the multiplicity of musical tones as *musica*.

For scholars at the time, the multiplicity of music was that of a flock, a chorus, a discontinuous gathering whose identity could shift over time. Like the stars in the heavens, the multiples of tones that made up musica obeyed laws (mathematical laws) but were neither infinite nor abstract. Unlike magnitudes, multiples could not be divided without losing their identity, yet it was only through the multiple that the ideal essence could be perceived. A tone described a unit of one but was not in itself the source of concord, which was a relational attribute. These relations lie, as mentioned, in the ratios between tones, and such ratios were presumed to be rational. In a baroque process of abstracted division and mathematical speculation, the tone and its intervallic relations colonized the earth and cosmos alike, producing intervals so minute that they were beyond human hearing and planetary orbits of such magnitude and duration that only God could audit their infinite sounding. However, not all differences between tones could be so reduced, and the attempt to do so would produce a complexity that was beyond imagination and reason. In Plato's *Timaeus*, for instance, the interval of the fourth (4:3) is "filled up" by the demiurge, "with intervals of the tone (9:8)." Because the remaining interval, equal to 256:243, was obviously absurd, it was agreed to leave a "remainder" as dubbed by Plato. However, this remainder was made even more complex by the Pythagoreans—defined as a ratio of 531,441:524,288 and known as the "the Pythagorean comma" (Heller-Roazen 2011, 36). This comma was to become first the incommensurable, then the accidental, in an ontology which, although based on "the one," would still manifest in the multiple and therefore still be connected to the perceptible, the sensible. However, as Heller-Roazen shows, eventually this connection was severed. Inevitably, in the mathematical study of nature, "the ways of conscious perception and arithmetical consideration must part. ... To the perplexity of their contemporaries ... the disciples of Pythagoras resolved to follow an unexpected path: they renounced the evidence of their senses for the certainty of their arithmetic" (ibid., 37). This required not only that irrational ratios be awkwardly suppressed, but that the possibility of incommensurable relations, incommensurability as such, was to be concealed at all costs, since knowledge of the incommensurable demonstrated the limit of the theory of the unities, showing that "numbers cannot transcribe the measures of this world." The Pythagorean comma, signifying a pause, a brief silence, could no longer float in mathematical space as an unresolved incommensurable. Its existence would have to be kept secret, along with the fact that the world can only be mathematical by plugging your ears.

Being beyond discord or concord, the tone, in itself, prior to being defined as a multiple or a magnitude, is a concept outside of music, a sign with no referent, a relation that only roughly corresponded to an actual musical note. Similarly, in mathematics, the unit of counting is excluded from the assemblies of numbers. As Badiou emphasizes, "the one" exists only as an operation. "There is no one, only the count-as-one. The one being an operation, is never a presentation. ... The multiple is the regime of presentation; the one, in respect to the presentation, is an operational result; being is what presents [itself]. On this basis, being is neither one ... nor multiple" (Badiou 2007a, 24). The importance of this exclusion is not restricted to music and mathematics, as Western metaphysics has tied itself inextricably to the ontology of the one, where "being is subtracted from every count, [and is] heterogeneous to the opposition of the one and the multiple" (ibid., 26). Since "there are no mathematical objects," if Western ontology ties existence to the unit, it ties it to nothing. "Strictly speaking, mathematics presents nothing, without constituting for all that an empty game, because not having anything to present, besides the presentation itself—which is to say the multiple—and thereby never adopting the form of the object, such is certainly a condition of all discourse on being qua being" (ibid., 7).

The emptiness of this "empty game" is, however, of sufficient complexity to demarcate, divide, and distribute the world and its contents, first through cosmology, then theology, and, in the current era, politics. (This is particularly true when we consider the tone of voice later in this book.) What lies behind the multiple is an operative void, yet this void provides the only possible answer to the dilemma that the operation of the count presents: that the "one" is not one, but exists outside of ontology as something that defines being, but is not in itself being. The "one" is beyond presentation and outside of the one/multiple dyad, yet, and for this reason, "it founds, 'behind' its operation, the status of presentation—it is of the order of the multiple." Badiou argues that the dilemma of the void has been resolved by proposing that "ontology is not actually a situation ... that being cannot be signified within a structured multiple, and that only an experience situated beyond all structure will afford us an access to the veiling of beings' presence" (ibid., 27). Yet this solution only works within an abstract form of being that is beyond being, beyond presence, infinite, unknowable, and unpresentable. As such, it resonates with theistic notions of an "unmoved mover," and in particular, coincides with the paradox initiated by Christian Trinitarianism. As we shall see, the tone intermingles with and operates on behalf of religious and then later secular governance, in much the same way that the concept of tone defines musical relations and therefore what is music, but is not itself music. As sound, tone is always multiple, yet as pitch, tone is unitary; as sound wave, tone is continuous, while as frequency, it is intervallic and numerical. Aesthetically, tone seems to reinvigorate the romantic and sublime perception of music. Discursively, tone relies upon a distinction between periodic and aperiodic frequencies, musical signal and nonmusical noise, and this distinction is definitive in both music and information theory. Metaphysically, tone dissolves any neat distinction between particle and wave, instant and duration.

Musical tone, sonic tone (if that is not an oxymoron), and vocal tone are perfect examples of both the "majestic" and "subterranean" within the ontotheological spectrum, music being allied with the Infinite, the being beyond being, and the artistic sublime; sound, and in particular noise (both linguistic and aesthetic), being aligned with that which is not—neither sense, knowledge (rumor), nor being. For as I have argued elsewhere (Dyson 2009a), there is no "being" to sound insofar as there is no singular sound, no such thing as "a" sound, nor, beyond acoustical frequency, a unit of sound that could be counted in the ontology of "count-as-one." To hear "a" sound is to hear what cannot be, and such hearing would abandon the subject to eternal ephemerality, to a shadow existence of becoming rather than being, where the soul, according to Pappus, "wanders thereafter to and fro in the sea of non-identity, immersed in the stream of becoming and decay, where there is no standard of measurement."[1] And indeed, sound's phenomenal characteristics seem to support its membership in the void of unbeing, which, to extend Badiou's description, means that it is not only "neither local nor global, but scattered all over, nowhere and everywhere" (Badiou 2007a, 55). It literally has no point. Because the void is impossible to name, because it must surpass all being and be indiscernible as a term, "its inaugural appearance is a pure act of nomination." The name cannot name anything, it cannot "indicate that the void is this or that."[2] In sound and music, this dilemma manifests as a noisy or noise-ridden poetics (onomatopoeia, assonance, alliteration, rhythm and rhyming, slang, repetition, etc.) in speech, and in music as the doctrine, still pervasive, that music expresses nothing other than itself. Both are to an extent indescribable outside of the unitizing system of music and acoustic computation—both remain speechless.

Paradox and Theology

So great was the drive to eliminate aberrant factors in the gradual homogenization of the medieval universe that a disciple of Pythagoras, or so it was rumored, was drowned at sea for revealing the existence of the Pythagorean comma and therefore the possibility of incommensurability. This would seem a fairly drastic punishment were it not for the importance of denying incommensurability among medieval scholars, determined to install a more mathematically coherent system both within music and the cosmos. Regarding the world as made up of individual units, discarding the multiple in favor of the defined quantity had been in operation since the early Greeks—this in itself was not new. What was remarkable about the severity of the disciple's punishment was its materialization, if you like, of an even more severe operation that was taking place, one that would have equally murderous consequences. For what was being installed in silencing the disciple was a "working paradox," a form of "cognitive dissonance" that would abide by an ontology based on the paradox of count-as-one, and glorify the "mystery" of the three-in-one, while literally drowning the voice that uttered the possibility of incommensurability as such.

Accepting some inconsistencies and refusing others is an operation not of logic but of power, expressed in the developing scientism of the medieval world on the one hand, and the consolidation of theocratic rule on the other. Heller-Roazen best expresses the importance of this development:

> Pythagorean thinkers of the ancient and medieval periods had defined consonances by numbers because they believed such sounds to be, in essence, mathematical. But they never claimed as much for all physical beings. The ancient and medieval cosmologies, at least in their Aristotelian forms, were hierarchical in structure, and they distinguished between beings of many types. At the summit of the scale of Being lay the eternal spheres, whose perfectly circular movements could be precisely defined because they were mathematical. At the lowest level, there were corruptible bodies, in themselves inconstant and therefore uncertain. Between these two orders of Being, music effected a conjunction: thanks to the doctrine of harmony, eternal ratios could be perceived in sounds. In the age that began with Galilei, however, the

> grounds for that conjunction would give way. When the stratified cosmos of classical *epistēmē* came to be supplanted by the homogenous universe of modern science, no being, in itself, would be identifiable with a number or a relation of numbers.
>
> That principle finds a precise correlate in the early modern philosophy of mathematics. (Heller-Roazen 2011, 69)

Abandoning logic in the quantification of nature, abandoning the senses in the perception of tonal harmony, music theorists put in place a doxology that connected to and reinforced another major cognitive discipline: theology. Like music, the Trinitarian article of faith that installed paradox at the heart of (divine and nameable) being, and at the highest level of imagination (God, the Infinite, cosmic time and space) also grappled with the necessary multiplicity of what could only ever be defined as "one and indivisible," a problem that manifested in a form of nominalism—specifically, how to name the origin of being. The debate over tone as a multiplicity or a magnitude that troubled music is relevant here, since popular names such as chaos, flux, cosmos, universe, existence, and so on describe a multiplicity, and therefore accidental action, randomness, serendipity, and a summons or call to fate. In the contemporary debate, flux and chaos are both linked to emergent behaviors, and as I have argued elsewhere, subtly presuppose an order based in evolution that is manifest through modern technology. If, however, one decides to nominate this origin as "the universe" or God, then the specter of a unitary being presents itself. So while nomination is seen to be simply "a name," it is precisely the name that is important. Despite the fact that flux, for instance, is held as the multiple beyond all multiple, its singling out is also a making singular. As Badiou points out, "What privilege could a multiple possess such that it be designated as the multiple whose existence is inaugurally affirmed? ... This question is none other than the suture-to-being of a theory—axiomatically presented—of presentation" (Badiou 2007a, 67). In mathematics, it has been interpreted as the presentation of the multiple of nothing, designated by the old Scandinavian letter ⊠, emblem of the void, "zero affected by the barring of sense" (ibid., 69). This naming was, according to Badiou, a semireligious act:

> As if they were duly aware that in proclaiming that the void alone is ... they were touching upon some sacred region, itself liminal to language; as if thus, rivalling the theologians for whom the supreme being has been the proper name since long ago, yet opposing to the latter's promise of the One, and of Presence, the irrevocability of un-presentation and the un-being of the one, the mathematicians had to shelter their own audacity behind the character of a forgotten language. (ibid.)

It is not surprising that mathematics would attempt to rival the dominant theology while at the same time incorporating pagan mythology. In fact, however, the logic that enabled the representation of the void also operated, and had done so for centuries, in metaphysics and in music. The definition of a tone as a multiple meant that it could not lay claim to infinite division and therefore resisted the doctrine that nature itself was composed of numbers. Yet at the same time, there was sufficient abstraction in the concept of a tone to allow it to act as an operator—a relational sign rather than a material thing. The tone's unique relationship with sound set it apart from other forms of representation such as those used in arithmetic and science because, whereas the latter employs signs to signify material things (having substance, duration, extension, in short, physical materiality), the tone signifies a particular relation between things (notes) that are without substance or extension. The concept of tone, then, is a double abstraction: a sign signifying a "nonthing," or to be more precise, "nothing." In terms of everyday sound, this has resulted in a dominant metaphysics and popular worldview that privileges vision, marginalizes sound, and employs a vocabulary wherein sound remains at the edges of description, forced into the awkward and paradoxical mode of "poetic" (a.k.a. noisy) speech in language, the scientific and mathematically based description found in acoustics, or the pseudo mathematical science of music. As is often the case with sound, such marginalization works both ways, for in terms of musical sound, until fairly recently it has aligned the tone—representing immateriality, ephemeral reality, invisibility—with supra—being rather than nonbeing. It is no wonder then that music would be the phenomenon of choice in representing divinity in one case and the cosmos in the other—or more precisely, that music's relationship with the cosmos would very easily be mapped onto its later use in religious practice.

The cosmology that Pythagoras developed launched the monochord into the heavens, leading scholars to argue that the planets, given their geometrical rotations and relations, must, as moving magnitudes, also be sonorous since movement produces sound. For whom would they sound though? As the planets required an ear to hear them, a mind to perceive them, and a reason for their co-relation, they affirmed the existence of a divine being, an infinite universe, and the sanctity of what was called "the fourfold" all at once. Yet at the same time, these neat correlations instituted one of the major flaws of unitary thinking: the impossibility of change. For philosophers at that time, it was inconceivable that the supreme deity would be satisfied with the endless repetition of the monochord's limited tonal range; it was also inconceivable that the deity would not also find pleasurable the discordant tones sounding irrational numbers. There had to be a parallel between the human and godly sense of harmony and musical pleasure, and there also had to be some distance between God's listening and his creation. It was recognized that musical pleasure also consisted in surprise and delight, that a certain amount of dissonance is necessary for music to develop, to excite, to create a sense of expectation. In fact, it was suggested that such dissonance was most pleasurable when it veered only slightly from the consonant tone. But the question remained as to what kind of melody planetary movements would compose and how pleasing it would be to the ear of the Infinite. Celestial music introduced a dilemma between aesthetics and mathematics in that for music to be pleasing, a certain irregularity had to be present, but this very irregularity is at odds with the concept of divinity.

Heller-Roazen summarizes a debate between "Lady Geometry" and "Lady Mathematics" conducted by Oresme, where Lady Geometry concedes that "if the motions of the stars issued in music, they could never be commensurable, for then they would be unchanging. Eternal asymmetries would be more beautiful" (Heller-Roazen 2011, 57). The incommensurable was necessary for both human and divine musical pleasure, and it was tolerated as an adjunct, rather than a counter, to musical aesthetics. Not surprisingly, this tolerance continues to enliven modern music, despite various subcultural incursions.

Economies and Trinitarianism

> If we do not understand the very close connection that links *oikonomia* with providence, it is not possible to measure the novelty of Christian theology with regard to pagan mythology and "theology." Christian theology is not a "story about the gods"; it is *immediately* economy and providence, that is, an activity of self-revelation, government, and care of the world. The deity articulates itself into a Trinity, but this is not a "theogony" or a "mythology"; rather, it is an *oikonomia*, that is, at the same time, the articulation and administration of divine life, and the government of creatures. (Agamben 2011, 47)

Agamben's study of *oikonomia* is highly relevant to our present-day "economy," as the same processes that enabled incommensurability within Christianity to become established (albeit as a "mystery") within the *doxa* of the being of God also ratified the cosmic-theistic relation, based on number alone, that music had inaugurated. This study is based on the distinction between political theology, which "found the transcendence of sovereign power on the single God, and economic theology, which replaces this transcendence with the idea of an *oikonomia*" (ibid., 1). According to Agamben, the meaning of *oikonomia* developed from the "administration of the house [*oikos*]." This house should not be thought of in terms of the modern day single family home, but is rather "a complex organism composed of heterogeneous relations, entwined with each other [through economic relations that are] linked by a paradigm that we could define as administrative [*gestionale*], and not epistemic: in other words, it is a matter of an activity that is not bound to a system of rules and does not constitute a science in the proper sense" (ibid., 17).

The model of administration was extended through the Stoics, where it signified "a force that regulates and governs the whole from the inside [to providence]—providing for the needs of life, nourishing" (Agamben 2011, 19).[3] Thus a divine plan extended from the position and orbits of the planets to the eventual salvation of human souls. In the Christian era, *oikonomia* also becomes the *task* of implementing the divine plan, which is figured as a mystery. The mystery is that of the Trinity and the incarnation of the Son through a division of one God into three, which separates but does not install a difference between them.

In the writings of Paul, the mystery becomes the administration of the plan itself: "[In designating] the Trinitarian articulation of divine life [the] Pauline phrase 'the economy of the mystery' [is reversed] into 'the mystery of the economy' *oikonomia sacramentum*" (Agamben 2011, 38), which confers on economy all the semantic richness and ambiguity of a term that means, at the same time, oath, consecration, and mystery. "The concept of *oikonomia* is the strategic operator that, before the elaboration of appropriate philosophical vocabulary—which will take place over the course of the fourth and fifth centuries—allows a temporary reconciliation of the Trinity with the divine unity. In other words, the first articulation of the Trinitarian problem takes place in 'economical,' not metaphysico-theological terms" (ibid., 36).

Various models were employed to describe the three-in-one, possibly the most popular being music. For instance, according to the second-century theologian Tatian: "*Oikonomia* tends persistently to be identified with the harmonic composition of the threefold divine activity in a single 'symphony'" (Agamben 2011, 39). Just as a symphony must contain irrational proportions, so too the Trinity can appear as a division into three, and yet through its administration, which is itself a "mystery," maintain its unity. The motivation to represent music and *oikonomia* in mathematical terms was subtended by the debate between magnitude and multitude: for the division into three of the single entity, God must incorporate magnitude; for music to contain the incommensurables that pleasing harmony require, it must be multiple. The solution to this dilemma built on the logic already inherent in the concept, and strategic operation, of *oikonomia* and paralleled the shift in meaning of *oikonomia* as "economy of the mystery" to "mystery of the economy." Incommensurables were placed within geometry, which meant that they could be studied as *analogies*: "Once this principle had been accepted, the threat of the irrational was in part contained: the incommensurable could be defined as an object of geometry, and arithmetic could proceed undaunted in its consideration of multitudes, all commensurate by virtue of their definition as diverse collections of many ones" (Heller-Roazen 2011, 46).

Having the status of an analogy—a sign—rather than a substance, or thing, the tone became a nonmaterial element in the production of harmony. Music itself was now a result of the "mysterious" workings or arrangement (economy) of single notes, rather than the notes themselves. As magnitude, if only analogically, music was of the same order as the Trinity: its identity could be secured in both cosmic and theistic unity; its harmonic constitution paralleled the "economy" that enabled the Trinity to be rational, logical, and mysterious; and its analogical nature meant that the trope of musical harmony could be applied to other spheres without necessarily being backed by any phenomenal substance. The immateriality of music thus complements the necessary immateriality of the divine, while still inserting the power of agency working through its worldly manifestations. Music is known, in the same way that the Trinity is known, through the sacred.

By the early 1600s, Kepler had devised the exact harmonic ratios that the planets would sound according to the distances between them. For instance, "The extreme divergent intervals of Saturn and Jupiter make slightly more than the octave; and the converging, a mean between the major and minor sixths. The diverging extremes of Jupiter embrace approximately the double octave; and the converging, approximately the fifth and the octave. But the diverging extremes of Earth and Mars embrace somewhat more than the major sixth; the converging, and augmented fourth."[4] The "celestial harmonies," replete with alto, base, soprano, and tenor, would of course sing the praises of God, even though, as Kepler admits, "voices or sounds do not exist in the heavens on account of the very great tranquility of movements," and further, "there is no such cause in the heavens, as in human singing, for requiring a definite number of voices in order to make consonance." Despite these evident facts, Kepler nonetheless felt compelled to argue that some "mind" must be perceiving this glorious symphony.[5] For Kepler, the harmonies of planetary movement could only be intentional, part of a grand design: "It follows, I say, that he, the artisan of the celestial movements himself, should have co-joined to the five regular solids the harmonic ratios arising from the regular plane figures, and after both classes should have formed one most perfect archetype of the heavens."[6] Devoutly religious, Kepler no doubt modeled the position of the planets revolving around the sun not only on the positions that singers take in creating polyphonic harmonies, but on the "choir of angels" that in Christian theology existed in order to praise God. Aristotle had already made a connection between the order of the cosmos and the structure of a chorus, producing harmony, but only through the governance and direction of a leader.[7] However, Aristotle's formulation lacked the administrative status that, by the time of Kepler, was deeply connected to the mythology of angels. Whereas Aristotle's god was undivided between being and praxis, the "unmoved mover" simply existed, and from his nature all proceeded, the Christian God relied upon a series of divisions, first between Father and Son, and then between the different hierarchies of angels, sorted by their proximity to God. The angels have both administrative and mystical functions, and their sacred gradation, arranged and monitored by God, unites the "mystery" and "ministry" of what is essentially, according to Agamben, the activity of government, which as such implies an operation, a knowledge, and an "order [whose] origin and … archetype is the Trinitarian economy" (Agamben 2011, 153). It was, after all, the Trinitarian economy that began the duplication of functions, the process of separation within the one being between power and administration.

Aquinas worried about the logical consequences of angelic proliferation: in terms of quantity (How many thousands? How many thousands of thousands?); status (Which angels were closer to God? Which had access to the divine secrets? Which were "ministers" as opposed to "administrators"?); and the protocol of their sociality (Which angels could talk to other angels? Could a higher angel talk to a lower angel?). Yet this infinite division, so reminiscent of the Pythagorean ratio, was held together not by number alone, but through harmony and song, in particular, the Sanctus—the song of praise.[8] Kepler's arrangement of the planets into "voices" that could form a celestial chorus can be partly explained by the importance of the voice in praising God. And in the Sanctus itself, we have the vocal imitation of echo, defining both a space and some form of material existence, as if providing a sonorous glue, much like the ether imagined by the ancients and theosophists alike. The echo of the angels reaffirms the existence of God, just as the songs of worship on Earth reaffirm a connection to heavenly administration and the heavenly kingdom: "The Angels are the guarantors of the originary relation between the church and the political sphere, of the 'public' and 'politico-religious' character of worship that is celebrated both in the ecclesia and in the celestial city. [The 'public' here] is defined solely through the song of Praise" (Agamben 2011, 146). The multitude of voices sings into eternity, creating the space, through their resonance, of infinity and populating this space even after the Day of Judgment. The propagation of angelic function that so worried Aquinas serves another purpose after all: it enables the world to end with the salvation of the last soul, while still preserving divine power. By "separating the hierarchy from its function," the angelic hierarchy is thus able to outlive its function.

This separation, like all others we have been discussing, has a metaphysical role that is as relevant today as it was in the days of Aquinas. As Agamben points out, if the Trinitarian economy is or was "constitutively tied to the action of God and his practice of providential government of the world, how can one think of God as inoperative?" For this also involves "the impossibility of thinking what precedes the creation of the world or follows its end," and thus, thinking the finitude of divinity. The echo of the angels fills the void that had been left by the possibility of the Infinite; their calls or cries fill the heavenly spheres with sound, just as the slow movements of its the planets, in their silent sounding, had provided a symphony for God's ear. According to Agamben, in the replication of divine names in liturgy, "The ineffable God—in apparent contrast with his unsayability—[is] ceaselessly … celebrated and his praises sung. … Ineffable sovereignty is the hymnological and glorious aspect of power that … [is celebrated] by singing the Sanctus" (ibid., 155). The song of praise unites the celestial heavens with worldly governance, angelic choirs, and the hymns sung during the church service. The angels "spread" the glory of God through their song, which is also an echo ("holy, holy, holy,") that obliterates the void through the repetition, reverberation, and reflection that echoing suggests.

2 Acclamation

The paradox of glory, according to Agamben, is that, as integral to God's being as it is, glory can neither be increased nor diminished, yet glorification through praise is what all creatures owe to God and what he demands from them. "From this paradox follows another one, which theology pretends to present as the resolution of the former: glory, the hymn of praise that creatures owe to God, in reality derives from the very glory of God; it is nothing but the necessary response, almost the echo that the glory of God awakens in them" (Agamben 2011, 21). The problem of infinite praise also impacts the longevity of the *oikonomia*, for the point where music might cease to flow, where the accident might be repeated, where God, having heard all possible melodies, might become bored, or the angels, having saved all souls, are left with nothing to do, is the point when the *oikonomia* must come to an end. But how can divinity end? The *oikonomia* survives, according to Agamben, through the "song of praise" that forms the basis of earthly worship. Through liturgical practice, music that is inaudible except to beings that are unknowable is brought into the range of human hearing. Music lifts the voices in the choir toward the painted ceilings in Gothic churches, their ascending spires reaching upward just as the voices of the choir, placed above the congregation, are reflected acoustically, sounding in a place above the churchgoers head as if coming from the heavens. But, as Agamben suggests, the continual production of praise for a being who does not need it "betray[s] the attempt to explain the unexplainable, to hide something that would be too embarrassing to leave unexplained" (ibid., 224).

If the hierarchy of angels established a divine precedence for civil administration, the graphic representation of liturgical song had a similar function in the ontological hierarchy of sound and the senses. According to Heller-Roazen, the system of writing that emerged in the eleventh century not only established the basis for Western polyphony by allowing the simultaneous arrangement of several voices that while harmonized, still remained autonomous. More importantly, however,

> the transcriptions illustrated principles of not sound but its conception. ... The empirical reality of song was gradually "objectified" and "conceptually abstracted" by a graphic practice founded on five fundamental notions: "the perceptual and, later, quantitative concept of the interval [the tone]"; "the consequent notion of *discrete sound*"; the idea of "the height of sound" which permitted the classical distinction between the acuity and gravity of sounds to be spatially exhibited on the vertical axis of the page; "the concept of the *note*, resulting from the ... preceding conceptualizations"; and finally, "the notion of musical scale." These five concepts all rest on the principle that to be musically intelligible, sounds must be discrete in quantity, like the old multitudes of arithmetic.[1]

Developing their "graphic semiotics," medieval musicians also considered durations as well as intervals to be multitudes, based on fixed units of time, while Oresme also introduced timbre, based on the observation that sounds are different according to the materials that produce them.[2] From the materiality of animal guts to the inaudible movements of the celestial spheres, materiality is embedded in sound. Yet it is an inaudible materiality: in the same way that Pythagoras discounted the forge (the space, the smell, the thickness of the air, the sweat of the men, the reverberations of the bricks or stones that surround them), Oresme quickly ascends from the depths of intestinal actuality to speculations about inaudible heavenly movements. These, as we learn from Kepler, could be tabulated according to a detailed harmonics and geometric relations. The issue here isn't so much the development of geometrical relations in the acoustical sphere as the imperative to move very quickly from materiality to speculation, from the depths of the earth and the site of sound production to the infinite possibilities of movement and sound that are both cosmic and unknowable. This tendency is repeated throughout—sonorous materiality is aestheticized or deified. In both cases, it can remain unknowable, reaching into imperceptibility as a continuum of microtones, inexpressibility as its microtonal structure encompasses noise, and finally, unknowability as cosmic relations decree a grand design—be it theistic or chaotic.

The movement from forge to heavens, from animal intestines to the inaudible yet still sonorous pulse of the planets, is a familiar theme in the history and philosophy of sound. Whereas in *Sounding New Media* (Dyson 2009a) I mentioned the progression from anaerobic to anechoic with regard to the evisceration of the voice and its dematerialization in the cogito, in this context we can understand the tone as being built from the same conceptual foundations. The tone of the voice is therefore a double quantification: the structure by which "tone" is established and the structure by which tonal variation in the voice is quantified, especially via computational methods. The point here is not simply to make a correlation between "tone" and "voice," as the conceptual structures that enabled a correlation to be made in the first place emanate far beyond the range of voice or the perimeters of music. Anaerobic and anechoic control is as much spatial as acoustic. Thus it is no coincidence that from minute fluctuations in frequency and volume (very small space and very small sound), universal concepts and projections into the infinite seem to flow as if part of a logical formula. Yet the actual trajectory of this formula passes everywhere but through the physical reality that sponsored the first observation. Cutting across space, drawing coordinates, and imposing vertical and horizontal axes, capturing sound within the note, harmony within the concept of the tone, concordance and dissonance within the idea of tonal relationships, grand design and cosmic harmony within a silent—or at least inaudible—sonority, these are all strategies of dematerialization rather than desonorization, of escape from the sweat of industry and animals.

The task of avoiding the paradox presented by the infinite division and impossible arithmetic that the Pythagorean comma and Trinitarian multiplicity inaugurated is continued in modern philosophy, acoustics, and aesthetics via Leibniz (1646–1716), who in 1712 wrote:

> Music is an occult practice of arithmetic, in which the mind is unaware that it is counting. For in confused or insensible perceptions, the mind does many things that it cannot remark upon with clear apperception. In effect, one would be mistaken in thinking that nothing happens in the soul without the soul itself being aware of it. Therefore, even if the soul does not have the sensation that it is counting, it nonetheless feels the effects of this insensible calculation, that is, the pressure that results from the consonances, the displeasure from the dissonances.[3]

The soul had previously acted as the "instrument" of God in its role as the voice of the conscience (Aquinas) and as such was an internal channel for the divine. Under Leibniz, the soul was to act as a calculator and recorder, a mediator between arithmetic and the sensible, between number and aesthetics, silent abstraction and the resonant ear.[4] Importantly, Leibniz, influenced the shift from the unit of tone to the unit measurement (in vibration) that Galileo developed, allowed a place for incommensurability in his theory of mind by "acknowledg[ing] that an absolute pure tuning of all musical Intervals is impossible, and substituting instead "convenient equivalences." The rapid calculation that formed these equivalences also created a movement/transformation from the unconscious sensation to the conscious awareness of harmony. Furthermore, Leibniz suggests that incommensurable relations may actually produce pleasure: as Heller-Roazen writes, "When 'dissonances' come between consonances ... they are 'pleasing' by accident; when they are 'slightly distant from rational relations,' they are, moreover, regularly 'pleasing' in themselves" (Heller-Roazen 2011, 86). This pleasure, "impenetrable to our senses," indicates for Leibniz that a hidden harmony must obtain.

Despite the relative decline in liturgical ceremonies in the modern state, acclamation, inoperativity, and the question of finitude still function, perhaps even more so. The rituals of parliament, the cheers and standing ovations of Congress, and the liturgical responses of church services all involve a degree of incantation, of repeating a word or action that is little more than "an empty rotation, [but yet] it is precisely this empty turning that supplies it with its peculiar, almost magical, efficacy: that of producing glory."[5] The "amen"—an acclamation that is always in response to something said and always in agreement—is a form of echo, as is applause or shouting out "Bravo!" Agamben mentions Nazi Germany and the instigation of "Heil Hitler" as exemplary, but it is in the sphere of public opinion that acclamation has thrived:

> The sphere of glory ... does not disappear in modern democracies, but simply shifts to another area, that of public opinion. ... What was confined to the spheres of liturgy and ceremonials has become concentrated in the media and, at the same time, through them it spreads and penetrates at each moment into every area of society, both public and private. Contemporary democracy is a democracy that is entirely founded upon glory that is, on the efficacy of acclamation, multiplied and disseminated by the media beyond all imagination. (Agamben 2011, 276)[6]

There are several aspects to this multiplication and dissemination. First, to take one of the most notable examples, it is well known that Hitler used audio technology, loudspeakers, and architecture, just as he used voice, song, and music to create a semireligious atmosphere, in which, as Nancy describes, the crowd was "feeding on its own exultation."[7] In this context, amplifiers, used to build a sense of increasing volume and presence among the crowds at Hitler's speeches, together with the architecture of the public squares created a form of feedback, an echo chamber within which ideology could be reflected back and expanded in the process. Apart from German socialist song, which, as Nancy writes, involves a summons "to the most profound interiority, to 'sentiment' itself understood as collective and unique to a defined community" (Nancy 2007, 56), this escalation of ideology occurs through one aspect of sound—its volume—supported by a particular architecture of space and an infrastructure of amplification. At Nazi rallies, it was not just Hitler's voice that bounced around the public square, nor the crowd's acclamations, nor the slogans and songs, the patriotic anthems, the sound of feet stamping and hands clapping. It was also language wrapped in the flag, an exhortation to act in a particular way, enfolded within music on the one hand and the sound (artificial though it was) of massive support and agreement on the other: "amen." Nazism, Nancy reminds us, "did not come out of nothing. ... Something had already been preparing itself for a long time—something that did not as such prefigure the Third Reich, but that offered it a choice space" (ibid., 50–51).

The "choice space" that Nazism was able to occupy was in part a product of ancient ritual and the twentieth century's devotion to the birth of a new humanity, a "new man," as Badiou writes, who found expression in the arts. Badiou illustrates the idea of the "new man" via reference to Wagner's *Ring Cycle* and the periods of fascism that also saw acclamation resound again in the city squares. Indeed there are many correspondences between the musicality of national anthems and Wagnerian motifs, and these, as we have learned, share a commonality with political/theological economy. But what I want to concentrate on here is a different correspondence—that between so-called discordant music and the creation of a new kind of listening, technologically aided, that harmonized the clashes that technology brought. Within the triumphant harmonies and vocal exuberance of the *Ring Cycle*, within the loud bangs of industry forging metals into something else, the ringing cry is of an ideology calling for swords to be turned into ploughs, of an end to war through industry, and of an end to misery through consumption. Within the dissonant sounds that were the mark of twentieth-century music, the production of a certain kind of audition takes place, one that absorbs this ideology without being conscious of it. This is the expressiveness of music, not in the romantic sense (as expressing the emotions or the spirituality of the composer), but as expressing the desire of the age, the desires of those who would produce and listen to such harmonies, who would stage them, record them, broadcast them, set them in stone as statues of the masters of the twentieth century.

How does listening absorb an ideology? Peter Szendy—whose enlightening book *Listen* (Szendy 2008) details the production of listening in modernity—unites both the sense of an ideal music with the notion of clairaudience, advocated by Pierre Schaeffer and still influential today, via an analysis of Wagner's reading of Beethoven—in particular, his deafness. Not being able to hear becomes, for Wagner, the condition for an inner clairaudience, as it allows the composer to escape the noise of the world and become "all the more transparent to himself," assisting the composer to hear, as Wagner writes, "only the harmonies of his soul" (Szendy 2008,

69). Like Aquinas's instrument or Leibniz's calculator, Wagner's soul, left in peace, will transmit the profundity of existence. But as Szendy points out, to hear this profundity, the listener also has to assume a form of deafness. "To listen without any wandering, without ever letting oneself be distracted by 'the noises of life,' is that still listening?" (ibid., 48). The echoes of power, articulated by the walls of the public square, literally resonate through what Nancy describes as "the timbre of listening"[8] within the newly minted individual, for whom identity and purpose are to be found in interior spaces, and an inward, individualistic turn.

R. Murray Schafer used headphones to retrieve the inner and spiritual self from the noise of the world. Since Schafer's experimentations, however, headphone-listening culture, most often associated with youth culture, has undergone an ironic but also quite cynical transference. Schafer's headphones are now the ubiquitous earbuds—an outward symbol of another form of exultation: resounding not within mass congregations but within the ear of the listener and whatever social networks their musical choices might engage. This ear, listening most often to the compressed sound of the MP3, has, as Jonathan Sterne points out, already been idealized or "perceptually coded" such that it "contains within it some model or script for hearing an imagined, ideal auditor" (Sterne 2012, 2). This auditor has, before any actual listening, already been determined as one who doesn't hear perfectly and doesn't hear everything and whose listening is shaped as much by "perceptual technics"[9] as by the music he hears, purchases, and shares. Not being able to hear perfectly in this case allows the MP3 listener to participate in a global media culture organized around efficiency, mobility, and normalization, that extends music appreciation to the sound of compression itself:

> As with the quest for verisimilitude, compression practices have created new kinds of aesthetic experiences that come to be pleasurable in themselves for some audiences—from the distortion that is a side effect of electrical amplification in radio, phonography, and instrument amplification, to the imagined intimacy of the phone conversation, to the mash-ups that aestheticize the MP3 form and the distribution channels it travels. (ibid.)

For a generation that has been defined by media advertising as integrally, perhaps even innately connected to information technology (IT), such that their social life, education, and very identity seem to depend on the acquisition of a mobile media, the coincidence between youthful exuberance, consumer electronics, so-called cloud culture, and the number and frequency of economic bubbles that have risen and burst over the last decade provides an ironic if not tragic allegory. It could be argued that "i" (pod, phone, pad, etc.) culture created the soundtrack for the dot-com bubble and became the poster child for an industry that has single-handedly concealed the demise of the empire by manufacturing of a form of optimism—a hyperoptimism—associated in particular with youth.[10] (The irony is of course that this optimism fuelled the conditions under which today's youth would find that sunny attitude very difficult to sustain.) In light of the influence that this culture, founded on an individualized hearing, still exerts, the loudspeakers that Hitler used may appear as quaint metaphors. Yet there are strong parallels between the present (or recent) optimism and the exultation that Nancy refers to.

> This space of exultation, of highlighting and dramatizing, always harbours the most formidable of ambiguities. But it harbours this resource most dangerously exactly when it presents itself as, and when it sets out to be, expansion—outpouring, overflowing, dilation and sublimation, the propagation of a subjectivity. (Nancy 2007, 50–51)

The "propagation of subjectivity" associated with i-culture is also defined by a listening that binds the future and the possibility of a future to the temporality of sound. As Nancy writes, "[Music] never stops exposing the present to the immanence of the deferred presence, one that is more 'to come' than any 'future.' A presence that is not future, but merely promised, merely present because of its announcement, its prophecy in the instant. … Prophecy in the instant and of the instant" (ibid., 66). Music has a special relationship both to the present, and importantly, to the anticipation of the future, and to an expectation of existing (hearing) in the time of this future.[11] It is therefore the perfect accompaniment to the liturgical repetition ("holy, holy, holy") mentioned previously since the expansiveness of a simulated echo is also projected into an imagined future. The infinite space and time of the divine is reiterated through worship, as music softens and extends the echoing in the Sanctus, filling the space vacated by the choir after their final repetition, and thus deferring the possibility of finitude. [12] Similarly, in its potentially infinite resounding, media-echo divides and multiplies like the Pythagorean comma, producing thirty-second sound bites that end up as a back beat, and inserting tracking and monitoring devices to recreate meaning (commensurability) where it has evaporated. As Sterne surmises with respect to iPod users' listening experience,

> listening is diffuse, distracted, and interconnected. Music becomes a soundtrack to other activities, which can mean many different things. It can be an attempt to overcome alienation and boredom; it can render the strange familiar and the familiar

> strange; it can be a tool of solipsism and mood management. ... It can be a form of self-assertion against an indifferent world; or it can be a form of self-dissolution in a world that presses down upon the subject. (Sterne 2012, 236)

As we shall discuss later in the book, the diffusion that Sterne refers to is part of the process of division, separation and multiplication that manifests in fractured media and politics: as the repetitive, and often nonsensical "gibberish" (to borrow from Serres) of political speech, the fragmentation of financial entities, and the negative equations and formulae that constitute economic "growth." The other "eco"—ecology—is captured by modeling and simulation, casting the experience of climate change within the banks of big data, while the actuality of the weather, its day-to-day record-breaking presence, becomes catastrophe's alias, providing perfect storms to explain the often-methodical larceny that is occurring in these times of multiple "crises." It is only "on the ground," within the reverberative echoic space of bodies that this weather is experienced, that its furor is heard as ominous sound (the rumbling, the crash, the scream) rather than a potential item of news to be repeated on the hour, creating a highly mediated, controlled, and detached unity through disaster.

We are all familiar with the ritual of blanket broadcasting political speeches that interrupt network programming; the devotion to election cycles and celebrity news, and countless other examples of opinion making. Increasingly, media consumers are being called upon to respond with a modern-day "amen" through live responses to, for instance, talk-show debates, broadcast as tweets on the bottom of the screen in much the same manner as the stock ticker. While much has been made of the "Twittersphere" and the contribution that social media is making to, in particular, election campaigning, it is not just the technology that creates a contemporary version of the public rallies reminiscent of the '30s. It is also the demand for engagement, the assumption that an "opinion" about almost every topic under the sun is something that every individual should be eager to share.

We could extend Agamben's argument regarding acclamation to include contemporary media, with, for instance, every tweet, text message or notification on one's cell phone, a reaffirmation of belonging. Acclamation in this respect is participating in the simulation not of eternal life, but of eternal, technologically aided communication, of endless talk, through the machine, to the machine. This call and response, a kind of liturgy, transmitted through media could be thought of as the new Sanctus —the eternal life it supports being reconfigured as endless communication through which the operations of power are retained, while power itself relies upon a "structural impossibility, while simultaneously disavowing this impossibility"

(Agamben 2011, 230). Indeed, for Agamben, just as the "fracture" that the doctrine of the Trinity cannot resolve is, in early Christendom, concealed by glory, and expressed by doxology, the fractures in political and economic structures are concealed by ideology. The latter becomes "a symbolic field which contains such a filler holding the place of some structural impossibility, while simultaneously disavowing this impossibility" (ibid.). As Agamben demonstrates so thoroughly, liturgical song has been instrumental in the development of political theology, itself a forerunner of economic theology. Like its predecessor, economic theology could only develop through the use of complex forms of diction, which include amplification, distribution, modification, and the glorification of particular voices within a sea of "media" sound. While it's easy to grasp the former—historical documents and cinematic remakes are characterized by what would seem to the modern ear to be an overemphasis on the rituals that are accompanied by singing, call and response, and the obvious display of wealth—it is more difficult to understand or even hear what I am calling the "media-tone" of our times. This is a tone neither musical nor linguistic nor even strictly sonic, since it is constituted as much by images as sounds. But its omnipresence and surveillance capabilities create a background hum, an audio-visual ambience that has assumed a similar status to the once-theological attribution of these operations. The "look" and tone of, for instance, the politician's speech has assumed critical importance within the context of ambient media, as it can be registered in passing, its meaning transmitted within an instant. It operates as an iconic sound/image flashed in a glance, a form of shorthand delivery only possible within ubiquitous media. The model for this kind of audio-visual penetration is sound—more specifically, tone.

If we thought about a media-sound ecology, we would notice that certain frequencies, certain sounds, are dominant—that the contemporary soundscape has a recognizable tone, not the "hubbub" or "chatter" of the past but more like a "gabble," lacking the hum of hubbub, being individualized through audio and other filters to produce a unitary speech. Sometimes these voices will speak at once or will interrupt each other. But even so, their speech will be clear, relatively loud, sharply articulated, and to the point—especially in the context of news and current affairs. These are massive generalizations, but they describe the contemporary soundscape that is of direct interest to the average person, that concerns their current and future well-being (who will govern them, who will pay them, how much they will earn, what they cannot do, and so on) and that has a particular sound. Listening to its particular sound will tell us as much about the body politic as it would about the body of the individual speaker who has a lump in the throat or a head cold. The tone of the (media) voice is also the tone of culture heard not through millions of people speaking but rather through the voices that represent them and through the multiple and diverse systems of amplification and transmission that these voices use to create their own reverberations, their own resounding, and ultimately their political, cultural, and economic dominance. As a result of such dominance, certain interests govern, their voices become voices of interest; they receive attention from the listener and amplification from the transmitter. Their particular tone overwhelms the sonic ecology.[13]

Listening Post

Mark Hansen and Ben Rubin's *Listening Post*[14, 15] brings together the new doxology of technoromanticism with older rituals of liturgical chant.[16] There are two versions of *Listening Post*. The first, winner of the 2004 Golden Nica Prize from Ars Electronica, has been far more widely exhibited than the second, which has largely been relegated to YouTube clips preserved by fans. In 2002 (when the piece premiered), Facebook had not yet launched, the global financial crisis had not yet occurred, the phone hacking scandal had not riven *News of the World*, and cybercrime was not yet considered to be a threat within the homeland. Yet the second version of this artwork, produced at the same time as the first, predicts some of these events, casting a dark shadow over the bright future of a globally connected online community. I will deal with the first now and the second later in the book.

Listening Post has been described as a "performance,"[17] an "informational landscape,"[18] a "machine that speaks the world's thoughts,"[19] and an art installation that offers a "near religious experience." Rubin and Hansen introduce their piece in their article "Listening Post: Giving Voice to Online Communication," with the following:

> The advent of online communication has created a vast landscape of new spaces for public discourse: chat rooms, bulletin boards, and scores of other public on-line forums. While these spaces are public and social in their essence, the experience of "being in" such a space is silent and solitary. A participant in a chat room has limited sensory access to the collective "buzz" of that chat room or others nearby—the murmur of human contact that we hear naturally in a park, a plaza or a coffee shop is absent from the online experience. (Rubin and Hansen 2002)

This introduction also underscores the way that the piece configures sound, as a medium that can, almost magically, become part of the "being" of the Internet, the sensorium of online infrastructures, and the experience of engaging in real-time screen-based communications. In a move that has almost come to epitomize the rhetoric of new media, sound—a phenomenon requiring physical space, material objects, movement, and atmosphere—is joined with a concept of "space," which appends, without qualification, the architectural metaphors, the "chat rooms" of the Internet. This is a move that sound—materially and culturally—enables. In the same way that the voice passes through and between language, music, and noise, as I mentioned with regard to tone, sound moves between material and message, representation and thing. *Listening Post* exemplifies this subliminal movement as the spatiality of sound mixes with the fictional space of the Internet giving the latter a materiality, a medium through which text can be transformed into voice. The musical elements in the soundtrack invoke music's association with the sublime, the ethereal, and the unrepresentable, linking the melodic and tonal qualities of the voice to the conceit of a universal language. As the sonic architectures shift from noise to music, from babble to speech, from chatter to chant, the three types of sound used in the piece, click, tone, and voice create a triptych that situates Internet communication within three familiar and related cultural frames, the first being visual, the second religious, and the third musical. The piece moves within these frames, inhabiting and then deconstructing one after the other, extracting meaning and then destabilizing each in a nomadic occupation of the cybernetic imagination. In a strategy of evocation and withdrawal, approach and then evasion, the piece brilliantly encompasses a broad range of potent and grounding cultural narratives and orchestrates these into a narrative specific to the Internet.

The intelligibility of the numerous "readings" of Internet chatter fragments collected by software agents trawling the net was a major problem for Rubin and Hansen, and the model of the orchestra provided a perfect solution: voices would be pitched, their entry would be staggered, the sound would be emitted from multiple sources (and thus spatialized by the room), the fragments would be kept deliberately short, and the software design would allow for a growing chorus of voices based on the meaning or content of the individual fragments of text. Additionally, the text would appear on the forty small rectangular LED screens that form the material and visual part of the installation. Essentially, software agents trawl the web for conversations from Internet chat rooms and reroute them live to the listening post, where algorithms are in place to select preset word associations or phrases such as "I am" with these phrases as a starting point, adding successive iterations by collecting segments of chat that include these phrases and then adding them to the building chorus. As Rubin and Hansen noticed in their research for this project, the listeners' ability to understand the various streams of vocalized text was enhanced by "assigning different pitches to voices from different speakers" and using "monotone or chanting" voices to both separate voices within an installation space and provide a "more cohesive mix." In the piece, multiple voices are orchestrated into something akin to a liturgical chant in the same way that the multiple voices of a choir are unified by pitch and separated by harmony.[20] While the often meaningless content of the LED screens might be organized via musical structures, at the same time, its origins in the conversational mode of Internet chat rooms are retained: the piece still adheres to the common perception of the Internet as a democratic, secular forum, a part of everyday life, nonexclusive, nonhierarchical, and essentially multiple. Just as the ocean of discourse that constitutes the Internet is rendered meaningful through software agents that search out text corresponding to pre-given words and phrases (such as "I am"), the pitched voices and musical tones are "demusicalized" through the nonmusical sounds of clicking on the one hand and the computerized voice(s) that speak the text on the other. In this way, the synthesized voice evades adopting any kind of subject position.

It is worthwhile discussing the nature of these sounds. According to Hansen and Rubin, "The clicking of the relays [draws] attention to the location of the sound sources" and this engages a form of "everyday listening," while the overall soundscape, composed of "pitch, harmony and rhythm," engages the listener in a form of "musical listening." "Everyday listening" lacks the predictive power of musical listening because, within Western tonality, tones conform to patterns that are easily recognizable yet almost infinitely recombinant, and "musical meaning" (itself a highly contested assumption) is far less dependent on context, intention, and the reciprocal participation of speaker and listener that defines "conversation." The interplay between the everyday and the musical are essential to the dynamics of the piece. The everyday "click" casts a hail of small mechanical sounds over the installation space, and these function as a sonic surrogate for Internet chat, reanimating the conversational experience despite the fact that there are no actual voices in the piece. These mechanical clicking sounds harmonize (literally and figuratively) with the synthetic voice in the production of the familiar—and highly symbolic—musical form of the chant. The chant enables vast movements across fictional spaces and cultural conceits and brings into play the god's eye, panoptical and omnipresent view of the cathedral, and the democratic but also religious concept of an intermediary—of one voice speaking for many, and many voices sounding as one. And it does this within the frame of music—the organizing system for transforming sound and noise into an aesthetic form. Although the voice has a slightly British accent and is obviously male, it can only obliquely occupy the position of the priest—the singular voice that one would expect to hear surrounded by the multiple voices of the chant, the archetypical voice that intercepts and speaks for the multitude. In this way, the synthetic voice avoids potentially damaging associations with theistic dogmatism. However, there are other global, panoptic, and authoritarian eyes and voices that the piece must deter if it is to retain its conversational, democratic, cyberspatial atmosphere. The piece has been described, for instance, as "speaking the world's thoughts." What could "speak" the thoughts of another? What could, to use Hansen and Rubin's terminology, "sonify" the conversations that are rendered only in text, and in so doing, help explain the "patterns" and dynamics inherent in the multiple, simultaneous conversations on the Internet?

Of course, the goal of pattern recognition immediately raises the specter of Internet surveillance and tracking programs like Echelon, a point discussed by Margot Bouman in her article "The Machine Speaks the World's Thoughts," who convincingly uses *Listening Post* to critique the Orwellian model of total surveillance. As she writes, "The paradox that *Listening Post* brings out is that the more technologically advanced the computer search engine, or the greater the machine's success at retrieving global consciousness ... the more unmanageable it becomes. ... Meaning is imposed from without, with little relationship to the raw data it filters" (Bouman 2003, 120). But there are other ways that *Listening Post* intersects with the specter of surveillance. A listening post is, after all, a station for intercepting electronic communications, a position from which to listen or gather information, and more ominously, a point near an enemy's lines for detecting movements by sound. If not a functional surveillance system in itself, the piece can be seen as a compositional device for transforming the Orwellian connotations of intercepting and tracking Internet chatter—for many years now associated with terrorist activity—into the far more benign reading of a system that simply articulates, or sonifies, the thoughts of the world

Like any great piece of art, the echoes of these transformations are profound. Having listened to a voice that cannot speak, having heard fragments of private conversations through a robot guided by algorithms, the visitors to *Listening Post* leave the gallery with their beliefs reshuffled: in the midst of re-sounding the social contract and establishing a sense of quasi-spiritual belonging, the piece turns the Internet into a thinker, and surveillance into a priest. However, there is another rarely screened version of *Listening Post* in which the chatter of the multitude, now fully tethered to technoculture and finance, is reprocessed once again. *Listening Post* part 2 presents the other side of the theocratic, governmental, and financial coin, without the euphonic accoutrements of the liturgical apparatus (Rubin and Hansen 2003b).[21] Whereas in *Listening Post* part 1, we see the happy face of Internet surveillance—we hear music, a pleasant voice reminiscent of a BBC radio voice, and we hear the very positive entry point "I am"—in *Listening Post* part 2, we hear the whirring of the machine: the unrefined, monotone voice of an automaton, short snippets of boring statistics, financial information, often just gibberish. The voice in *Listening Post* part 2 is barely understandable: it is harsh, its words are harsh, it is both more sophisticated and more cynical than its predecessor. "Use your head," it says, in a flat, foreboding tone, filled with the hoarseness of transmission. It doesn't seem to be speaking to anyone and certainly isn't making any statements about itself, such as "I am." The music in the background, rather than uniting fragments of text and speech, pulls them apart, emptying them into a chaotic well of dissonance. The sounds are sharp, percussive, not necessarily loud, but disorganized, accompanying the fragments of decipherable speech in fits and starts as if the sound/music, the not-yet-music, were still in-formation, not yet information. It is a difficult piece to interpret: the raw, sharp commands of unfiltered power punctuating the dissonant soundtrack could be emotionless commands that cross the trading-room floor, or quite possibly the gibberish of a malfunctioning AI system. It is also literally difficult to hear because in both form and content, what we hear are filters-in-development, prefilters perhaps, that have let the raw material out before it has been properly refined and regulated. We don't get a sense of a world being spoken as much as an echo chamber, a voice multiplied, enclosed within a box, occasionally being lucid but more often than not spouting gibberish. Without the symbolic transformations of *Listening Post* part 1, the piece presents nothing more than the inner workings of a different kind of box, whirring and churning, grumbling and muttering, a system without a narrative. The tone of *Listening Post* part 2 is prescient, as if it issues simultaneously both a truth and a warning, revealing the subterranean echo that travels with the visitor after her experience of *Listening Post* part 1.

3 Infinite—Noise

As we have seen, dissonance, or "noise" in music, was a necessary although not always welcome component of the celestial harmony that liturgy enacted. However, there is another sense in which noise, dissonance, and discord contribute to the liturgical function in recent history. Although liturgy seems to have become less popular in worship, Kantorowicz points to an unexpected rise in the 1920s, revived by "theologians and musicologists at precisely the moment in which the European political scene was dominated by the emergence of totalitarian regimes." (Agamben 2011, 196). This revival also coincided with the reemergence of Pythagorean cosmology in theosophy and its uptake in experimental music and the music of the avant-garde. According to Douglas Kahn,

> Followers of the occult in the latter half of the nineteenth century took Pythagorean and Platonic thought to heart, as such followers do to this day. Whatever the reason they might have felt out of place, they continued to seek assistance from music to place them in the cosmos and the social order. Like earlier Pythagoreans, they were required to respond to criticism. Developments in mathematics and measurement made it obvious that Pythagorean notions were no more than hopeful arithmetic and geometry. Concerns in Western art music challenged harmonic traditions with increasing dissonance. And audiophonic technologies (telegraphy, telephony, phonography, radiophony) helped transform the nature of auditory culture itself, raised expectations that phenomena should be audible, increased the sphere of what had yet to be heard, and based all sound, music included, more firmly in a materialist rhetoric of acoustical vibration. (Kahn 2004, 108)

Pythagorean doctrines were, according to Kahn, associated with a movement away from the musical canon and the incorporation of extra musical elements, as a way of breathing life into the overly technical Western art music of the late classical era (although, of course, it is possible to hear in late classical composers not only elements of noise but the errant divergence of unorthodox rhythms, key changes, instrumentation, and progressions in the sonata form). Composers began to introduce and incorporate sound of their own particular zeitgeist, and sound itself began to substitute for the vast cosmos that had been lost in the drive to mathematize music. The mimetic birdsong in Debussy, industrial sounds in Stravinsky, sirens in Varèse, environmental ambience in Cage, not to mention the "art of noise" advocated by Luigi Russolo and the Futurists, demonstrate an expanding horizon of sonorous material that embellished and sometimes replaced musical (if pitched) content. However, tonality was also extended intramusically through the concept of the microtone and the shift from a tonal system to one based on frequency alone that occurred with the rise of electronic music. The emphasis on frequency and vibration obliterated the distinction between musical and nonmusical sounds and, with it, the argument for harmony as inherently rational.

Standing as an allegory to the necessity of change, the disharmony in music—which could also be seen as a disruption—provided an aesthetic demonstration of the necessity for discord that could very easily be mapped onto modernity itself. (Neo)romanticism and what were then new technologies brought the rhythms and roars of industry to the concert hall. Discord became the sound of change, and its proponents argued vigorously that harsh sounds and displeasing tonal combinations were essential for musical and social progress. In a form of neo- Pythagoreanism, the concept of progress became historically "ordained" by both nature and human purpose, by scientific-mathematical reasoning, and the threat that stasis presented.[1] Despite its sound and the uproar it caused, the movement of music was tied to the movement of man: necessary, natural, and beautiful. In art, philosophy, and culture toward the latter part of the twentieth century, the Pythagorean comma becomes the aesthetic embodiment of change that, like the slow-moving but sonorous heavens, is perhaps necessarily beyond reach: caught not within the planetary orbits of the ancients but within the imperceptible microtonal changes that place music within an epistemology of code and listening within the technics of audio. Far from being transgressive, then, noise, disharmony, and discord were recognized, if not as a necessary evil, then certainly as a necessary annoyance. In the twentieth century, the discursive reins were lifted, so to speak, from dissonance; atonality was now worthy of discussion, musical and acoustic dissonances could be theorized as a necessary component of the aesthetic and the widely held belief that noise is not music was put to rest. It is not my intention to disparage the use of noise in music—indeed, some of the most interesting developments in sound art are occurring through investigations into noise. However, it is important to situate the artistic use of noise in the latter part of the twentieth century, within the system of regeneration-through-appropriation that defines the shape of popular culture in the West. For instance, Jacques Attali argued some twenty years ago that noise and modern music heralded social change, opening formality to difference, the concert hall to the outside world. Not just composers, but writers, poets, and artists found in noise a source of inspiration. Henri Michaux's writing in *Space Displaced* is exemplary in the way that it literally draws parameters around sound and music, the center and the periphery, dominance and colonized cultures (Michaux 1992). Michaux describes his experiments with a musical instrument from Africa that he affectionately referred to as "sans" (the instrument that he was originally looking for was something small that wouldn't disturb the neighbors). He had neglected it for several years until one day, "sick of everything, laid up after an accident, my foot in plaster … useless again and stupid in my thoughts," he tried playing the instrument, with its broken body complementing the writer's own. The first sound that the instrument emits is already an impacted noise, which Michaux describes on different occasions as a bird call, a desolate cry, the voice of wisdom, despondency, brutal frankness, and an existential profundity[2] anchored by its anteriority to both music and discourse: "It was not discourse. Nothing was joining. It was simply denial upon denial. One single harsh and hostile note. And that sufficed. There was nothing in this of singing, everything in it the violence to song and the magic of song. Refusal, flat refusal, a brutal rejection of the agreeableness which is always there, of the conversations almost inevitably made when notes are prolonged to put together in a composition" (Michaux 1992, 67). The rawness of the sound, beyond language or music, nonetheless expresses protest and disdain, being "against the trivializing of sentiment," which for Michaux is the production of music itself. For the cry that is exposed is immediately withdrawn and stifled (like the last cry of "the disobedient urchin on being caught in possession of his theft") (ibid., 75). Muffled by power, this sound is transformed, through metal, into a fecund noise, "transmitted" by the raw material, the "cursed steel" that, through labor and industry, has produced the instrument that when whole and complete would emit mellifluous sound. In all cases broken, the instrument, its sound, and the performer expose the material transformations that have to occur in order for music to be music, and this exposition provides new material for creating "outsider" art.

United by accident, fate, and disability, the "sans" and author produce revelatory and revolutionary noise "in league against harmony, faith and hope and reason" (Michaux 1992, 71). If anthropomorphized, Michaux's musical horizon places the poor and underprivileged on the outskirts of musicality, not through intention—as would be the case with atonal composition—but through the broken body of the instrument itself, its inability to produce the sound its material body was intended to create. For not only is the instrument broken, it is discovered among poverty and desolation. According to Peter Brooke, who introduces the text, the specificities of the instrument's failings—its broken reeds, for example—relate to the attributes of rebellion, and describes them as "individualistic outlaws not keen to blend into some easy harmony or common purpose, but pulling in different directions" (Brooke, in Michaux 1992, 19). Here, in the context of poetry, noise is "an immeasurable anti-word: a sub-language never to be articulated or systematized, always on the verge, unexpended in its vibrant, though mutilated, virtuality" (ibid.). It functions as another form of plenitude, a form of anti-music, a kind of sonic limit defining the edges of the social. Noise plays the anthem of the underprivileged, united in their cacophony, patriotic in their rebellion, coalescing on the fringes of a society where "good music" like "proper language" comes from bodies that are whole, intact, doing what they're supposed to be doing, fulfilling an intention. The throttled cry (throttled by whom?) has the last word in a text. It exerts its power of presence and is only "half" imprisoned by summoning the power of magic. Noise, flux, plenitude, magic—contained like antimatter—are there on the outskirts, ready to supply.

The Big Bang

Despite its ubiquity, noise is one of the most difficult sonic categories in sound studies—difficult because its negativity hides an existential necessity. In a cosmic sense, that is, when we are speaking about the Infinite, noise is the background chaos or flux from which all being originates. For such a degraded sound, the eschatological range of noise is extraordinary, participating in origins and endings across millennia. Like the Infinite, like the concept of existence itself, it is impossible to imagine the absence of noise. As a pair, however, noise and the Infinite are impacted: from philosophy to aesthetics to cultural commentary, we find the sonic or static-filled (im)materiality of noise fusing with the metaphysical, cosmological, and eschatological abstraction of the Infinite to form an intensely productive but also highly unstable entity. Think, for instance, of the big bang—an origin myth that encapsulates some of the current dimensions of noise. According to some, the term big bang was coined by Fred Hoyle satirically as a way of emphasizing the absurdity of the theory itself. While Hoyle has denied it, the anecdote itself demonstrates the confused beginnings of this sound—which is at once an echo of a bang and the bang itself. It was noise, as static and interference, that first prompted astronomers to theorize a beginning to the universe (in the form of a massive, explosive, event), evidenced by traces of noise (static) that continue to reverberate through the galaxies billions of years later. Like so many instances of a noise that is simultaneously momentous and banal, this cosmic noise was at first seen as an interruption, something that stood between the signal and the information it held. In fact, the researchers were looking for ways to eliminate it. This is not surprising since the elimination of noise is integral to the concept of noise itself: noise being that which we don't want to hear. Yet traces of the big bang fall into a different category of noise because, paradoxically, these still-reverberating echoes of cosmic static establish a limit to the Infinite: through noise, the universe gets a beginning—the Infinite must now incorporate at least one aspect of finitude, of a "before" if not an "after."

The fact that the term "big bang" has endured demonstrates both the common desire for an origin and the ease with which such an origin could be bound to an explosion of immense sonic and volumetric proportions. In terms of sound, noise, and silence, the big bang united volume with largess—since the initial "bang" produced the space and time that would house an infinitely expanding universe. There could be no end to this productive trajectory, or certainly no end that could be measured. If the universe was so generous in its constant and unsolicited expansion, surely limitlessness could be imagined as a universal law. The "bang" gave a particular sort of sound—one produced through a series of interactions that culminate in a split-second release of energy—the imprimatur of an ancient and unimaginable creation. There could have been no ear to hear that original, although easily imaginable, explosion, and so, almost logically, before the bang one imagines there could only have been silence. "Silence" is a perfect correlate to nonexistence, being a "something" from which "something" can arise, yet also a phenomenon that is impossible to imagine, let alone hear. Used as a term to denote the absence of sound, the concept of silence represents absence itself, a paradoxical absence that relies for its meaning on an eternal presence. "Silence" is also, therefore, a term that conjures creation, production, in short, existence. The concept of the big bang introduces silence as an always, already existent from which existence arises via negation, enabling the universe to have always existed, and to come into existence at the same time.

Culturally, this origin paradox, like its theistic and philosophical predecessors, enables other paradoxes to sit comfortably within the modern imaginary and, until the last few decades at least, be expressed in the assumption of eternal growth. The short-term shortsightedness of this assumption appears, ironically, in the word itself. Perhaps Hoyle should have taken a lesson from the angelic choir, given that the onomatopoeic expressivity of "bang" rapidly diminishes after its initial vocalization. Or perhaps he was cognizant of the irony and contradictions that noise contains—for how can we talk about it without sounding like idiots? Who would in all earnestness argue that infinity itself could arise from such ambiguity, such insufficiency, such a ludicrous and demeaning concept, as if the human intellect could discover the origins of the universe but was incapable of articulating it without lapsing into infantile and incoherent speech? How could anyone have not understood the term as joke? But misunderstand they did, and as a result, the modern understanding of the origins of the universe echo, however faintly, with satire. One could almost say that satire reverberates through noise once it is taken seriously—once it is capitalized, once a "big bang" becomes "The Big Bang."

There are lessons in this capitalization. For here, along with a singular origin, is a singular sound. The explosive associations of "bang" delimit a beginning and an end, a sonic event rather than a sonic continuum. The "bang" is a noise among an overall noisiness, an identifiable sonic "thing" or "event" or even "object" that stands out, protrudes into materiality, and turns noise—the generalized hum that barely enters language as a category of the sensible—into sound. In doing so, sound is lifted from its multiple and continuous ephemerality to become "a" sound. Multiplicity and indifference (and by indifference I mean both a lack of differentiation and an attitudinal stance) give way to identity and interest. Difference becomes interesting just as identity begins to orbit not just the universe but also "our" planet, and ultimately ourselves. Noise as a singularity, perhaps "the" singularity, bestows upon the inhabitants of the universe the possibility of identity, individuality, and, ultimately, the ability to hear and speak of existence itself. In short, it bestows the Infinite upon, or within, the individual. Not "soul"—for we are far too modern for that—but inalienable rights, constitutional prerogatives, individual freedoms, and a degree of immortality, even if only as indestructible matter. With Noise (capitalized) as ultimate guarantor, it is possible to avoid the thought of finitude or the concept of a limit while at the same time drawing boundaries around the self that are strong enough to make individual rights plausible. Indeed, it was only on the basis of establishing "a continuous hierarchy of intelligences, angels, demons and souls ... [that] communicated in a 'Great Chain' which begins and ends with the One," that it was possible to construct either the unified subject of modern science" or civil administration and government.[3] In this context, Noise follows the long list of sonorous phenomena setting the individual within a structure of governance that entails an ordering so complex that it is virtually unquestionable. Through noise, the individual or group demarcates a space for itself as an identity, having a unique "voice" that can and must be heard through the din of doxa. The inherent "challenge" of noise has been drafted during periods of both musical and social experimentation, fuelling debate between traditional musicologists and those with a more experimental bent, between a culture of symphonies and a subculture of local live music.

Punk rock, for instance, barged its way through the polite discourse of musicology, tying noise to anarchism and its revolt to "no future." The appropriation of the more radical and political aspects of punk rock has been well documented, but what was interesting in the initial phase was the way that noise was so closely tied to the perception of decline: noise was in fact the expression of youth anger at the obvious end of growth. Only later did noise, then, as an initial reaction, a shouted response, a loud protest, find its way into popular culture and the art gallery as an aesthetic form and material in itself, "noise" becoming "Noise."

There is more to the ascendance of noise within aesthetics, however, than simply the imperialism of music and culture, or the necessary rebellion of new generations of artists against older forms. The entry of noise into music in the twentieth century can be seen in a much broader sociopolitical context that spans civilization itself, as a realignment of the tripartite division between consonance, dissonance, and the pleasurable, paradoxical, accidental, and amongst divine power, angelic administration, and the realpolitik of governance. It is not so much that each element gains or loses power, but that its operative mechanism (as dissonance, accident, exteriority, difference, interference, noise, the unaccountable, and the unrepresentable) come to the fore in persuading common sense and public opinion to accept a new reality. While the individual will, for instance, have access to legislation regarding, for instance, the noise level of the neighbor's apartment or the presence of heavy machinery in their street, they will have little control over the space that noise, in the name of industry, appropriates and lays waste to. Public opinion, as formed by popular talk shows and manifest as chatter, back chat, outcry, and so on, may be concerned with individual rights at the minutest level and will hold within its multiple voices conflicting views and ethical contradictions at almost every point of civil society, such that this discourse, chatter, or hum drowns out almost all other issues and concerns. Yet the global consequences of urbanized, industrialized culture are almost impossible to articulate within this scenario, since the appropriation of space also impacts sense such that, as Nancy writes: "All space of sense is common space (hence all space is common space). ... The political is the place of the in-common as such" (Nancy 1997, 88).

Noise serves many purposes, as do the concepts of sound and silence. They are all, to an extent, overdetermined and under-thought, but like fossilized remains rattling about the recesses of language, they open a perspective, an attitude of curiosity toward the concept of the "individual," who, despite Galileo, is still regarded as the center of the universe. As Hillel Schwartz in his groundbreaking book *Making Noise* writes: "Noise is never so much a question of the intensity of sound as the intensity of relationships: between the deep past, past, and present, imagined or experienced; between one generation and the next, gods or mortals; between country and city, urb and suburb; between one class and another; between the sexes; between Neanderthals and other humans" (Schwartz 2011, 21). I would add that of these relationships, possibly the most intense are those embedded in the concepts of sound, silence, language, voice, ambience, atmosphere, tone, and attunement, those that fluctuate between the sonic and the metaphoric, between sound as substance and sound as imagination. This is a "between" that noise negotiates, yet because it bears the weight of not only ontological but all logical wisdom (why things are, and the way things go), noise also registers the paradoxes, the tautologies, the spiraling catch-22s, the absurdities, the nonsense and gibberish that swamp our cognitive habitat. The more material noise proliferates, the more difficult it becomes to ward off noise's existential consequences—by way of description, aestheticization, and categorization. To say that noise can't be thought is, after all, not saying much.

These negotiations are not trivial, for noise figured as the Infinite penetrates the minutest of sound: tearing it from music, extracting it from speech, inserting a question mark over music and language, that is itself entangled in the impossibility of thinking "Noise." Whereas "Noise" suggests a multiple that can never be divided, a difference that can never be assimilated, "noise" confronts us with finitude, and its confrontation is radical. There is a kind of mutant paradoxical play, a devious irrationality, a recalcitrant refusal, a stubborn insistence on staying put that follows noise whenever it is led through the gates of musical aestheticization or metaphysical appropriation. Noise threatens acts of appropriation while at the same time its use as a material can be seen as appropriative. As a structural device—in philosophical schema, for instance—Noise operates below the radar of opposites and can function as an anchor of sorts. This fantastic malleability arises in part from the fact that while much can be said about Noise, very little of what is said can be definitive. In much the same way as the origin of the universe housed in a big bang resists interrogation, Noise as a conceptual model also has nowhere to go. Beginning with a bang, noise moves from undifferentiation to singularity, as a trace of an event that could only be singular. But before the bang? Silence? But it is a "silence" that can only ever be conceptual, an aural metaphor that relieves the thinker from the mental strain of thinking nothingness. Cosmic noise, by revealing the essentially conceptual and nonsonic essence of silence, introduces one of the more awkward relationships between categories of sound. We could think of this aspect of relations between sonic categories as almost an undoing or reversal—as silence becomes "nonsound" or "unsound," calling into question what would normally be considered a main vector in the sonic axis of sound, silence, noise, and music.

In terms of theorizing noise, it seems to me that cosmic noise places a limit on the process of deterritorialization and, by implication, calls into question at least one aspect of sonorous becoming within a Deleuzian ontology—an ontology, or more correctly a mode of philosophizing, that is often referred to in discussions of noise.

Nancy is one of the few French philosophers who has actually entered into a discourse on sound rather than appropriate sonic metaphors to substantiate other claims. He describes Deleuze's philosophy as one of "nomination and not of discourse. It is a matter of naming the forces, the moments and the configurations ... not to signify things but rather to index by means of proper names the elements of the virtual universe." The intention, or if not the intention, at least one of the consequences of such nomination is, according to Nancy, "to have language bear the weight of what it is not. The incorporeal laden with the corporeal: not by giving it sense or manifesting its sense, but by effectuating it differently" (Nancy 1996, 111). One of the central tasks in engaging with the discourse of sound, is to understand the fundamental problems of nomination that have, throughout philosophical thinking, rendered sound, outside of music, almost impossible to theorize.[4] The dual meanings of "noise," as both sonic material and signal interference, are both examples of the interwovenness and omnipresence of noise, its intractable multiplicity and conceptual permeability, and at the same time, an instance of the failure of language—the failure of (visualist) culture, to recognize the sonic and its ephemeral attributes. The process of nomination that Nancy refers to can apply equally to Holt as it does to Deleuze, in that the fricative "bang" must bear the weight of the concept of infinity. An impossible word bearing the weight of an impossible concept, muting forever the exclamatory gasps or exasperated sighs that indicate a blockage or limit—in this case, a limit to thought.

This question of nomination is addressed in another context by Michael Gallope, who demonstrates the relationship between the refrain in the philosophical sense and in the musical sense: "Metaphysically speaking, the refrains of life mimic [co-extensively and immediately] musical refrains" (Gallope 2010, 87). Through the adoption of musical terms and within the overriding musical metaphor, certain musical qualities are privileged. Rhythm, for instance, is concerned with the genuine production of difference—it is not simply "vulgar" repetition (Deleuze and Guattari 1987, 314). However, these are always relative terms; there is no exactitude with meter either. But the two terms certainly have different connotations in culture—rhythm being far more popular than the more restrained and perhaps academic or musicological term "meter." Rather than discuss the difference between the two as a function of time and periodicity, it seems more productive to concentrate on their difference in terms of reception at the cultural level. What such a reading suggests is that the terminology Deleuze uses is chosen on the basis of its impact as prose rather than its metaphysical meaning, providing an elaborate rhetorical architecture used to buttress what are—as Gallope points out—ethical or aesthetic judgments. Perhaps all philosophy is just a matter of rhetoric anyway, but it points to a difference between Deleuze and Serres at the level of rhetoric, the production of the text, the style of writing. Deleuze uses the musical metaphor, adopts the musical analogy in a baroque style of ever-expanding metaphoricity: words added to words, terms altered, reconfigured in a form of linguistic filigree. The musical models Deleuze has chosen in *The Fold*, for instance, follow the historical progress of Western tonality so closely as to make it virtually impossible not to map one upon the other.[5] In this way, a terminology that is peculiar to Deleuze blends seamlessly with musical terminology, and in this sense, the fold becomes more than a philosophical operative; it also describes the way in which Deleuze produces the text, as ever more complex iterations of the basic theme occupy increasingly large parts of the page. Serres's language, on the other hand, goes in the opposite direction, becoming sparser, more brutal, more noisy, becoming in other words the very thing he is describing. Here noise, the noise of the text, the sonic noise that Serres is referring to, is a way of grinding down and spitting out language, a way of bringing language back, in a style reminiscent of Antonin Artaud, to its "carrion" nature—not, as Artaud would have suggested, to its intrinsic nature, but to the way that it has become a form of gibberish as it circulates ever more abstractly in the toxic stratosphere that now circles the globe.

Because noise figures so prominently and problematically in metaphysics, because its incorporation into Western art (music) has also been a strategy for music to regenerate its essentially conservative institutionality, and because the aestheticization of noise is so closely tied to consumer culture, I believe it is important to question its resistive power—as both material and metaphor. As the various interpretations of Michaux's "Broken Music" reveal, noise is not just radical; it can also be grotesque. Ignoring noise as pollution also banishes the ugly, banal, unaesthetic, sonic residues of often-toxic processes from critical reflection. Here, noise is "unsound." It is uncontained and contaminating; its ephemerality leaves residues in the thickened folds of the ear, which are eventually blocked with earplugs or refashioned with hearing aids. Its immateriality is suffused with electromagnetism, nanoparticles, other "nonobjects" that, lacking the innocence of sound or the structural incoherence of noise, draw the sonic away from its privileged status as deterritorializer of note. Also, because noise infiltrates—it passes through filters in a manner that is both subversive and at times violent, we can expect it to disturb the theoretical boundaries that have been set in place—we can expect it to produce paradox, to disturb our values, to be both unsightly and to some extent inaudible. We hear noise as a mélange and a multiplicity, as a chaos and flux from which nothing can be discerned. Within the sonic arts, noise sits comfortably within our pseudo-theocratic, hyper-optimistic, futuristic culture as a domain from which no meaning need be extracted, analyzed, rationalized, or accounted for. One would have thought that noise would be less easy to assimilate, but information theory and a fascination with the "glitch," the residue, the excess, has resurrected noise. The sonic plenitude that noise represents in sound art is often associated with the seemingly unlimited potential of digital audio; yet, as Sterne writes:

> Contemporary writers tend to associate this age of composition with the rise of digital audio, the boom in sound art, the growth of sampling and recombinant music, and the wave of music piracy online and on the streets of many major cities. … [However] digital audio … remains firmly rooted in the ideologies and practices of several communications industries. Digital recording technologies may do just as much to standardize the sound of music … as to challenge those standards. The boom in sound art challenges the revered status of the museum, the musician, and the composer but it also upholds those notions, if only in the negative.[6]

In the work of experimental composer/sound artist Paul DeMarinis and the subgenre of experimental music known as "glitch," for instance, the sound of technology itself creates the sonic event. By using signal noise, amplifying background noise, attenuating the sound of the recording apparatus or the playback system, deliberately using the "rejects" of a work, glitch aestheticizes the sound of the technology playing the sound of the technology—that is, the technology sounding itself. According to DeMarinis, whose work predates glitch by decades, "when you play a digital CD, you don't hear music, you hear a CD."[7] Glitch would seem to be an involution, a negative, abysmal loop, a form of technological and sonic navel-gazing: the aesthetic object is not so much sonic as an index of its technologizing. However, it is the very paradox of listening to the sound of sound's production that, in its obscure eclecticism, its hyper-implosive iteration of the sound of technology (rather than the sound that technology is supposed to produce or reproduce) and its preference for the detritus, the waste matter of audio production, that lends this genre a proximity to noise. For, like noise, what can be said of music that produces itself as residue and waste? How do we relate to it as an aesthetic object when it's so obviously concerned with sonic content usually subtracted from the aesthetic object—the subtraction of which in part determines the shape of the aesthetic object?

According to Sterne, the aestheticization of noise runs parallel to the industrial use of noise to mask or even perceptually eliminate the noise of transmission in telephonic transmission:

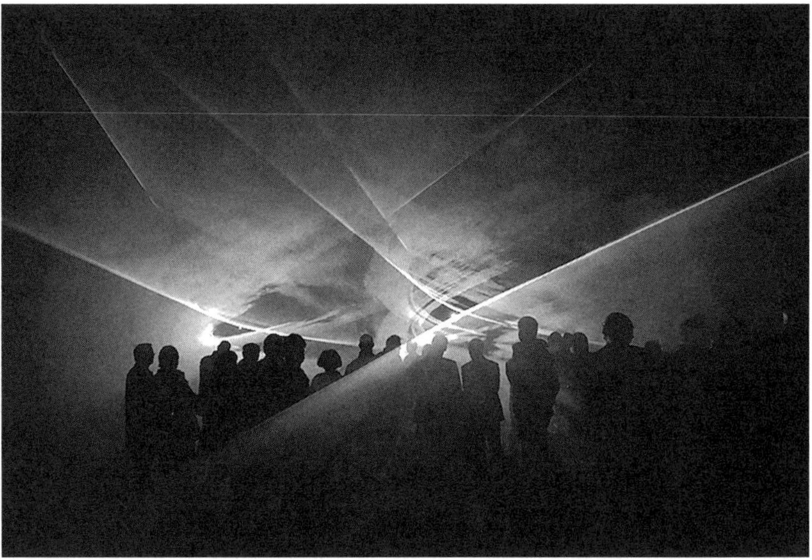

Figure 3.1
Studio Edwin van der Heide: *Laser/Sound/Performance*, Avantgarde Tirol, Lichtakademie Bartenbach, Aldrans, Austria, 2008.

> [In the '60s and '70s] while thinkers in a wide range of fields had defined noise as something to be eliminated, dentists, architectural acousticians, artists, and musicians began to explore ways of rendering noise useful. [Similarly] where computers had been used to model perception and as musical instruments, during the 1970s they were increasingly reconceived as sound media in their own right.[8]

This connection between industrial imperatives and artistic practices is most evident in the corporate sponsorship of art projects that would integrate experimental music and sound, noise as a sonic component, the engineer-as-artist, and IT research within the architecture of developing consumerism—be it a gallery, exhibition, event, or festival—such as the infamous 9 Evenings: Theatre and Engineering held in New York in 1966 and sponsored by Bell Labs, or, perhaps the most potent example, the Pepsi Pavilion at the Osaka world's fair.[9]

Toneworks: Tudor, van der Heide, and Ikeda

Attali's celebration of Noise as the harbinger of social change in 1977 coincided with a new attitude toward noise in art, engineering, and (some parts) of culture.[10] However, as David Tudor's piece for 9 Evenings, *Bandoneon!*, reveals, there was far more to noise as musical material than simply negative connotations. There are interesting parallels with the impetus behind Tudor's piece and the contemporary media artist Edwin van der Heide's *Laser Sound Projection* (*LSP*)[11] and Ryoji Ikeda's *Datamatics*,[12] with respect to their mutual fascination with tone and light, synesthesia, and the aesthetic experience of noise or tone pushed to the nth degree. Edwin van der Heide makes use of the canonic form and Pythagorean lines in order to introduce a process of incorporation and reflection within the spectator that puts the infinite multiplication and division of both Trinitarianism and Pythagorean doctrine into relief. Van der Heide's *Laser Sound Projections'* use of single tones to produce laser patterns in an improvisatory performance context is similar, conceptually, to David Tudor's experiments with sound and laser projection in the 1960s, and his tonal compositions could be read as an intensification and elaboration of some of the noise-paradox that had obviously influenced Tudor's investigations.

Tudor had experimented with sound and light some years before *Audio/Video/Laser* in *Bandoneon!*, performed during the 9 Evenings festival. Tudor described the piece in the 9 Evenings catalog as: "A combine, incorporating programmed audio circuits, moving loudspeakers, TV images and lighting, instrumentally excited. … *Bandoneon!* uses no composing means, since when activated it composes itself out of its own composite instrumental nature" (Hultén and Königsberg 1966). Tudor used the bandoneon—relative of the accordion—to produce two or three tones that could trigger visual images and control both the acoustic and luminous properties of the space. In doing so, he opened up a new territory in sound art—pushing the concept of tone to its limits (white noise) and using this sonic signal to trigger light waves projected on a screen during the performance. But while adventurous in terms of sound/light integration, Tudor's interest in the piece had as much to do with the modulation of tone and its transformation into signal as with the synchronization of sound and light. In a brief and obscure note regarding the work found in the archives of Fred Waldhauer, Tudor offered some clues for understanding the piece: "Pre- and post operative note—*Bandoneon!* sound/image is attending toward total oscillation—approaching white noise, it is differentiation discoverable therein ... it's theatre, performer activating interacting media, investigate." (Waldhauer 1966). A chief investigator in this instance was the Armory itself. Waldhauer—responsible for designing the "Proportional Control System," a sixteen-channel system that controlled the intensity of light and sound on stage remotely—recalls:

> As David played a certain note, for example, one light would become brighter and dimmer in response to the volume of the tone generated. Another note would change the sound level of one of a dozen altered bandoneon signals in a similar fashion. David began the piece on a sort of drone which turned on one of the lights, at first dimly, then brighter as the sound came up. Sound came through one of the twelve balcony loudspeakers. David added other tones and the armory responded with *its* sound.[13]

These elements—the transformation of tone into noise, of sound into light, and of the built environment into a dynamic, resonant space—are also present in *LSP*. This is not to suggest that van der Heide is channeling Tudor but simply to point to their shared fascination with the way these elements interact. Tudor's interest in white noise, feedback mechanisms, and self-generating compositions allowed him to explore the sonic and also cognitive complexities such dynamism presented. With reference to the enigmatic "pre- and postoperative" note above, if the sound/image is approaching a total oscillation and if its differentiation is "discoverable therein," then the discernment of tone or difference within noise turns the latter into musical and performative material. At the same time, the "sound" from the instrument also abandons its significance as "sound." Pursuing a trajectory of "sound-as-signal," Tudor's piece demonstrated a formalism that was both aleatory and technically circumscribed. In this way, Tudor differed from his longtime collaborator John Cage, whose piece for 9 Evenings, *Variations VII*, created the form if not the content of an aural narrative, supported sonically by found sounds that referenced the outside world and perceptually by fact that both sound and the listener's experience occur through time.

Van der Heide also eschews narrative in favor of a compositional strategy that, as he says, is based on his belief that "music is abstract—it doesn't need representation."[14] *LSP* begins innocently enough with a deep tone sounding in the recesses of the exhibition space becoming faintly audible as the visual perspective suggested by the laser beams shifts, like searchlights mounted on a horizontal surface that suddenly tilts. The background tone becomes clearer, approaches a recognizable pitch—a D perhaps—and as it reaches that almost musical identity, the laser beams curve, resembling sound waves, their blue, misty luminescence tinged with a cinematic green hue. The original tone is joined by a perfect fourth, forming a basic harmony that, in its familiarity, is barely noticed as it settles quietly within the ear, just as the mist settles almost imperceptibly upon the skin. But not for long—as these comfortable territories are interrupted by a discordant sixth, pushing into the space like an alarm: a constant beep, a siren that begins to rise in a glissando as the space itself seems to expand, radiating color and texture both within its walls and through the shining laser lights. Yet this glissando doesn't develop within itself—as it could, following a tonal line that aurally compliments the diagonal of the laser beam. As if reaching a point of premature exhaustion, it fractures, dividing and multiplying, scattering into a series of insistent beats that flash down on the spectator. The single tone, acoustically padded with near-pitch overtones, has become an audible and visible signal: a series of sounding/lightning flashes that capture and illuminate the room with their dancing diagonals, their geometric arabesques, pulsing out a rhythm (1, 2, 3, 4) that accelerates and again fractures and scatters, dissolving into an impossibly fast articulated drone—a tone divided within itself, inseparable from the light that is now flashing at the speed of a blur. This new tone, or drone, becomes the basis for another series of divisions, multiplicities, departures, and reunifications in another form, "like a dandelion," van der Heide has suggested.[15] A very slight, syncopated, high-pitched, and noisy beep, sounding like the overtones of a machine in overdrive, begins to lay its rhythmic pattern over an already-accelerated mix—itself the progeny of a single tone divided. We are now in an atmosphere that feels frantic and electric. An audio-visual, spatialized, electrified, glissando, a siren of light, sound, and fog that seems to reach its natural maturation as it explodes into static. Imagine two giant sparklers in a room and you have some idea of the sound and light energy that is filling the space.

The play between shadow and light, between clear tones and harmonic relations that then "fall" into disharmony, progresses via an elaboration of the harmonic intervals that are present in a harmonic, although inaudible form in any division of the scale. There is a Baroque elaboration, a fractal generation of the simple sounding of harmonic intervals—initially between the tonic and the fifth. It is as if an entire symphony could be created simply from this minimalist gesture: where a theme becomes a subtheme, melody becomes baseline, harmony expands and then it is reduced through microtonal expansion creating a wall of sound, a drone rather than tone. The same logic applies to the shadows and light that fall upon the spectator. Diagonals divide into herringbone patterns, their sharp lines curve and bend, creating an audio-luminescent fugue that grows in richness as tones divide, harmonize, develop discordance, fracture again, and shift from music to signal, tone to beep, beep to drone, and then, drone to tone. As sound and light transform effortlessly from the diagonal planes and clear tones that describe a "mathematical" space, to the swelling, swirling sounds and patterns found in electromagnetic-atmospheric phenomena, moving along the spectrum from pure tone to the "total oscillation" Tudor identified, music becomes signal, and tone becomes a series of beeps. But noise/static, we learn, is not the apocalyptic finale of van der Heide's piece, nor is the glissando that rises beyond human audition. Instead, the original tone returns, the laser beams tighten again into diagonal beams, the beeps merge into a drone, and from there return again to a single tone.

How does van der Heide move so eloquently between these sonic extremes? Why does he lead the listener through dramatic shifts in perspective, challenging their perceptual acumen while stimulating their senses? From aural tone and laser line to complex microtonal bursts (a.k.a. noise) and near stroboscopic luminescent pulsations (a.k.a. static) and then back again, all within the moist, atmospheric immersion of fog, van der Heide pushes the limits of perception yet not in the characteristically sensationalist modes of digital media; not to say "wait, there's more" in the blue-sky horizon of the digital (virtual) age. His approach is more complex, for the compositional trajectories that he traces illuminate both the past and the present, challenging the audience's understanding of their experience, inserting a limit to the limitlessness that lies at the rhetorical, mythological heart of past harmony and present virtuality. Tone and line: that perfect union of sensorial proportion that brings symmetry, order, and harmony to the world. Yet, as mentioned, always within this Pythagorean configuration there has been room for, indeed the necessity for, accident, discord, and the incommensurable. Leibniz best articulated the view that it is the accidental, the slight movement away from harmony that allowed God to accept incommensurability. The impossible ratios of Pythagoras could only be rationalized within Western harmony through this secret delight, this slight veering away from the tonal path, rationalistically ordained through vibration and mathematics. It could only be pleasing to the divine ear because, as Agamben points out, divinity itself must contain some irrationality, or incommensurability, or paradox, and this, he argues, lies at the heart of Judeo-Christian Trinitarianism, resolved only by an "administration of the ineffable" that could mediate between experience and the rational cogito. This system of separation and administration flows through the history of Western metaphysics, penetrating the individual as Descartes's theory of perception names the senses mere "transducers," mediating between the ineffability of experience and the certainty of reason. It could be argued that the perceptual limits of *LSP* also administer the categories of "noise," "tone," and "static," ensuring they remain intact—not substantively, but simply as a cultural shorthand for phenomena that fall into the "too hard" basket—beyond perception and immediate understanding.

What then does *LSP* confront its audience with? The limits of senses that are immersed and then assaulted, from all sides; the limits of bodies that, confronted with the simultaneity of sensual experience, only hear "noise" relative to the reduced tonality it first encountered; the constraints of a hearing that only grasps the beginnings (the perfect fifth) and fantasizes an end (the harmony of the spheres) of Western tonality; the bonds of theocracy that attributes such musicality to a global, universal compact, governed by the benign, although an indifferent hand of an unmoved mover. The quiet, seemingly ordered progression *LSP* establishes is suddenly hijacked by the too-rapid pulse, the too-complex microtonal mix of a tonal system accelerated beyond recognition. While sound becomes light becoming fog that, settling on the skin, lends itself to touch, while lines of light form atmospheric circulations, and sounds wash over and through the folds of the listener's ears, clothes, and ambulations, an electromechanical pulse brings the experience back to Earth, modulating the entire sound palette into a nontonal, although not necessarily atonal, musicality. Here we have something recognizable—music again.

With this return, *LSP* distinguishes his "beams of light" from an earlier composer, and one of van der Heide's influences, Edgard Varèse, who wrote, in his now famous essay "The Liberation of Sound":

> We actually have three dimensions in music: horizontal, vertical, and dynamic swelling or decreasing. I shall add a fourth, sound projection—the feeling that sound is leaving us with no hope of being reflected back, a feeling akin to that aroused by a beam of light sent forth by a powerful searchlight—for the ear as for the eye, the sense of projection, of a journey into space. (Varèse 1967, 197)

Varèse's description of sound as a nonreflective "projection" assumes an infinite and unpopulated space—a space without resonance or reverberation—and in many ways reveals the dominance of an essentially visualist, modernist, disembodied, and cinematic point of view. From the projector tucked away in a vestibule at the back of the theater to the all-encompassing screen, light moves through not just a volumetric, atmospheric space, but also through the mythology of the cinema with its enforced darkness and quiet. Be silent, sit still, contain the flow of the voice, breath, and limbs that would otherwise jump, gasp, and sigh at the moving configurations of light that the viewers perceive in front of them. Such stillness and quiet are testaments to the strength of the cinematic *projection.* A physical beam of light transports the view, and worldview, of modernity, reaching out into the future to speculate about the three-dimensional and immersive cinema in, for instance, *The Shape of Things to Come* that H. G. Wells described as early as 1933, cinematically realized in the film *Things to Come* in 1936. For Varèse, the idea of sound projection influenced his unfinished work, *L'Astronomer,* in which the protagonist is "volatilized" in outer space, disappearing with no hope of return in a melancholic sacrifice to scientific progress. For Wells, the projection would become holographic, ubiquitous, propagandistic, and a homogenizing force in the highly disciplined monoculture of 2030. *LSP* could be seen as one manifestation of the cinema's speculations, offering the immersive experience of cinema's "new media" offspring, dissolving the screen, the separation, indeed the medium between arts or performance and its audience. However, no one falls to the ground in mystical rapture, for this experience is given to us within a particular context: a gallery, a venue, a room that can contain what would once have been experiences of the miraculous. It is given to us within the context that has "normalized" the enhancement of perception, the augmented reality, the curious manipulation of belief that accompanies immersive experience, where our sensations are bracketed by the caveat "digital" or "virtual." In the gallery, it might seem for a moment that we can have it all, that we can safely hold on to this "god's-eye" view while still being fully immersed—rational but speechless, sensorially vibrant in the here and now but with a premonitory eye to the future. But this experience is reduced through its enclosure, just as van der Heide necessarily restricts the range of tones (frequencies) used to produce audio/visual correlations to its harmonic basis, the bare bones of Western tonality from which other tones and melodies can be imagined.[16]

Similarly, Ryoji Ikeda's *Datamatics*[17] offers a perfect representation of the vertigo produced by attempting to comprehend the Infinite within the infinitesimal, to appreciate the graphic vectors while simultaneously hearing and feeling the deep audio drone that seems to bind these charts, graphs, and symbols, describing the cosmos and genome in the same visual gesture, the same sonic breath, for the viewer's aesthetic and sensual experience. The piece was performed at the Festival of Music and Art in Hobart, Tasmania, in January 2012, and a later installation-based version of the piece, *data.matrix [n°1–10], 2006–09* (2009) was also exhibited at the Museum of Contemporary Art Tokyo in 2012 at the "Searching for a New Synesthesia" exhibition. The performance in Tasmania begins with sounds and numbers flickering on a giant screen like television snow. A light tone, constituted by a range of frequencies yet carrying a recognizable pitch plays from multiple speakers in the surround-sound auditorium—a converted warehouse on the port. The "static" on the screen is actually a grid of very small, rapidly changing numbers moving through some sort of numerical sequence at a speed that gives them the shimmering quality associated with television snow. The multiple tone(s), on the other hand, reminds one of the sea—in terms of its soft, white noise; a convocation or a crowd at a stadium—in terms of its echoing multiplicity—or, alternatively, the first few minutes of the film score, where the music or sound accompanying the titles begins to blend with opening shots of an environment—fields of long grass rolling gently in the breeze. Gradually, a low frequency tone is added, so low that the listeners can feel it rumbling in their stomachs, alerting them to their internal state of being. Quite suddenly, this tone is pierced by a high-frequency, very high-pitched single tone that lacks the harmonic texture of the other sounds. It arrives as an imposition, a signal, a voice ordering the sonic landscape, establishing sonic parameters, a high and low, thick and thin, signal and sound, instruction and reception. Suddenly the cohesion provided by the deep, vibrating tones, the rough but still easy-to-listen-to static, and the wash of what sounds like multiple voices, are broken. The signal tone punctuates the noisy but familiar warmth of the sunny environment that has been building, growing in complexity and volume, yet after a minute or so it too begins to blend with the mix; even the thin, horizontal red line that appears on the screen intersecting the static is assimilated.

Thus far the audience is still swathed in ambience—conversations proceed uninterrupted, people rearrange the seating on the floor, get up to go to the bar, arrive and sit down, greet each other, carry on socially. As if to order this ambient mix of voices and objects brushing past and engaging with each other, a sharp, quite shocking beep introduces a pulse—the soft world of noise is now lost to tone and a pulsating, beeping rhythm.

So begins the suite of Ikeda's sonic/visual variations on what could be described as the theme of ordering. Each deals in some way with the visualization, sonification, and quantification of massive amounts of data representing unimaginable complexity (for instance, a genetic sequence) and the infinite vastness of the universe. Each piece begins fairly innocently: this screen is black with small white dots resembling the stars on a clear night, the entire panorama rotating slowly to the soft, harmonically complex noise. This time, the beep has a clear visual analogue—a red crosshair that "captures" each dot/star, naming it Ross 148, Alpha Sentry.

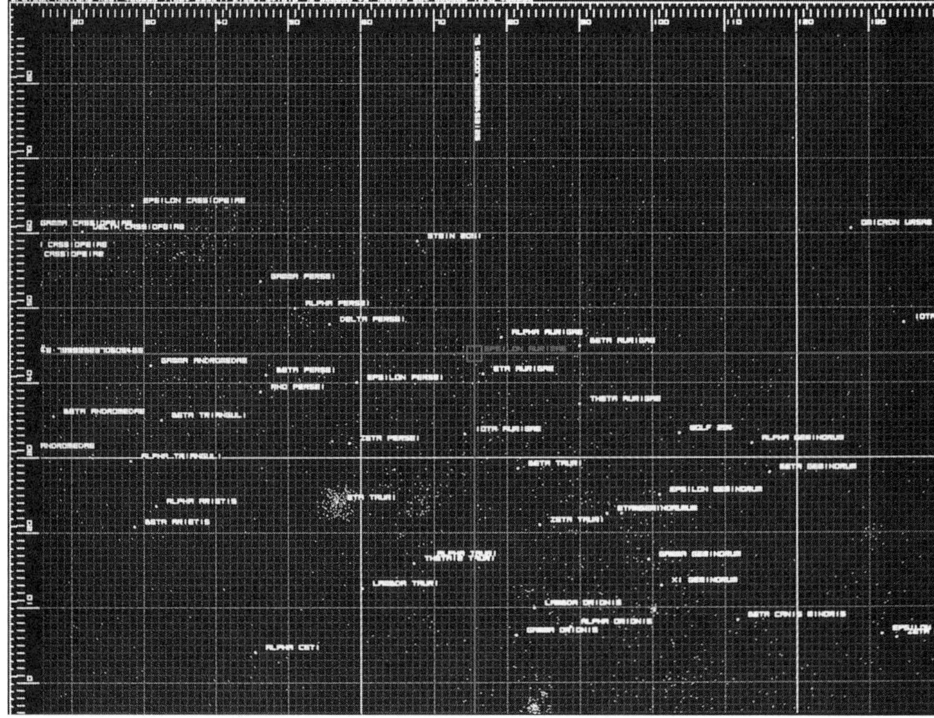

Figure 3.2
Datamatics (prototype ver. 2.0), 2006–. © Ryoji Ikeda. Photo by Ryuichi Maruo. Courtesy of Yamaguchi Center for Arts and Media (YCAM).

The audience realizes that stars in the galaxy are being targeted, named, and given numeric coordinates, as letters and numbers are punched out, bright white typography associated with computer printouts, the familiar typography of data. A three-dimensional grid forms, the targeting becomes more encompassing as the beeping and high-pitched tones proliferate. Eventually, the entire screen is filled with the names of different stars. The tones change, the sonic atmosphere becomes more frenetic, and the names and coordinates are supplemented with a readout of their magnitude, galactic positioning, and other information that passes so quickly on the screen that the eye barely has time to register it. Information flashes in the same way that the beeps sound—not for the ear or the eye to understand or appreciate, since the sound is harsh and the visual information flashes too quickly to be understood, the names of the stars disappear before they can be deciphered.

Ikeda's visual patterns move from signal to noise, literally and metaphorically, moving in and out of visual and aural perceptibility, as, especially in *Event Space*,[18] the rapidity of visual pattern and the high pitch of the tones produces a dizzying sensation. Noise—both visual and sonic—engulfed by, perhaps returning to its roots in, nausea. If this noise is evidence of the limit of human perception, so too its corollary in eco/logic evinces the limits of comprehension. As we will later discuss, while it is commonly accepted that measuring both the market and the climate is a necessarily reductive enterprise—otherwise economists and climate scientists would be overwhelmed with data—the setting for this reduction is, like the gallery, also a setup—enabling the simulation of experience or understanding only because of the walls that the gallery, or Wall Street, provide.

4 Disaffected Voices

The big bang not only announced that the universe erupted from noise, but its adoption as a quasi-tongue-in-cheek term became the rallying cry of twentieth-century moderns: "God is dead."[1] The mathematization of music, the musicalization of sound, and the aestheticization of noise could, at one level, offer examples of the desacralization of the divine and the end of acclamation. There can be no devotion to the big bang because its onomatopoeic name and association with noise already foreclose the possibility that it could take the place of what used to be called the cosmos. But we could also pose this question in reverse. If "God is dead," who, or what, is calling? Who, or what, is praising?

Part of this reconfiguration involves not just the object of the acclamation but the acclaimer. While angels—without actual voice or body—may no longer ring out eternal praise, other "posthuman" bodies have taken their place. The quantification of vocal tone, the synthesizing of the voice, and the development of human-computer interfaces based on human-machine conversation, create agents that provide the "who" in this post-theological era. If the correlation between God and human no longer operates, perhaps it has been replaced by a techno-gnosticism that, in its mishearing and miscommunication, and particularly in its methods of representing and quantifying the tone of the voice, reveals the absence lying at the heart of this new relationship between humans and machines.

Both a difference in degree and a difference in kind, posthumanism signifies an extension of the technological logic of modernity, characterized by the drive to increase efficiency, speed, acceleration, productivity, and, ultimately, economic growth, as well as a momentous shift in our understanding of what it means to be human, brought about through the emergence of complex systems that blur the boundaries between the organic and the machinic, the actual and the electronically mediated.[2] Notions of place, community, self, reality: these are at stake in the transformation underway. Yet, although posthumanism might simply involve a rethinking of what constitutes reality, the self, or the other, there is also a sense in which the magnitude of technological change is not restricted to mere notions, but, through developments in AI and ubiquitous computing, is reshaping the bodies, spaces, and objects that constitute our very materiality. If such developments constitute not only unprecedented acceleration in our technological immersion but an absolutely different, unique, shift in the conditions of human existence, then it is worth taking seriously notions of the self, community, and reality that are being transformed. And if, as Timothy Lenoir and others have argued, "we need not simply acquiesce to a view of the posthuman as an apocalyptic erasure of human subjectivity, for the posthuman can be made to stand for a positive partnership between nature, humans, and intelligent machines" (Lenoir 2002b, 211), we need to ask how this partnership might manifest in our *actual* exchanges with intelligent machines.

One way of approaching this question is to examine the emergence of a system, device, space, network, computational agent, or artificial entity that sheds light on what it is to be posthuman. The artificial voice, currently in the process of being generated by autonomous, independent agents, is a very good candidate for such as investigation. Having been amplified, transmitted, recorded, and distorted for decades, the voice is now heavily (though not necessarily successfully) involved in the development of bots, robots, and software agents that not only speak but converse. The artificial, distributed voice raises questions that directly pertain to the possibility of technological agency: we might wonder, for instance, what happens to the autonomy of the agent if it is ventriloquizing single or multiple collections of human voices. Or how does the substitution of human for machine voices—now favored in interface agent design—rebound upon the ideal of intelligent agents co-creating our social and material environment?[3] This chapter will focus on the moment of interaction between human and agent—a moment that, as I argue, is modulated by the aural and the social in ways that introduce an unanticipated, unaccountable, and chaotic element (a reverse-emergence, if you like) into those systems—technological, epistemological, and discursive—that together produce the artificial voice.

The smart homes, virtual workplaces, automated call centers, and various environments in which autonomous agents compliment or substitute for human face-to-face conversations are shaped by the technological interface, the spatiotemporal coordinates, and the quality of human engagement. While the latter can be assessed retrospectively through user feedback and proactively through interface design, the actual moment of interaction between human and agent is fleeting, ephemeral, and influenced by a plethora of factors that designers could never anticipate and users would rarely recount. Like the voice itself, conversation disappears into the temporality of sound, becoming an experience, an event that escapes complete recollection or interpretation. The experience of this event is, as Jacques Derrida argues in *Echographies of Television*, in essence the experience of the other, which arrives without expectation, without assimilation, and speaks with its own voice (Derrida and Stiegler 2002, 102).[4] Pursuing Derrida's line of thought here, how do we experience the autonomous agent: where does it come from, to whom does it belong, and in the context of human–computer conversation, who is it addressing?

Similarly, taking the metaphor of *conversation* as a goal for human-computer interface design seriously (as do researchers in the field), how can the particularities associated with human-computer converse, and the questions of agency that ubiquitous computing raises, be brought into focus? Paying attention to the *tone* of interactions with autonomous agents, to the way that human subjectivity is transformed by the texture, the tenor, and the rhythm of these acoustic, linguistic, and communicational events, reveals much about the transformation of self, experience, and sociality that is underway. For in many ways, these aural attributes, determining the contours of the conversation, grounded in and by the voice, both support and destabilize the hyphen that sometimes separates human and posthuman. Moving between established and new forms and technologies of telepresence and simulation, resisting codification, the voice engages with autonomous agents in ways that trouble the clear transmission of both meaning and intention. In the following I will discuss this dual and sometimes contradictory movement of the voice: between speakers (human and posthuman) and technologies (analog, digital, media, and informational), and between the smooth-flowing and awkwardly interrupted conversational moments that symbolize our somewhat ambivalent relationships with intelligent machines. I will do this by tracing the voice through two unlikely conversations—one between Derrida and Bernard Stiegler, and the other between an embodied conversational agent and its human client.

Voice

Because of its long record of both reproduction and transmission (through, for instance, phonography and telephony), the voice has anchored the presence-at-a-distance that electronic media activate, establishing a familiar, nonthreatening avenue toward technologies that might otherwise seem to disembody the user and atomize the social. The easy transplantation from human speaker to graphic simulation (for instance in cartoons) has provided an embodied model to which the simulation can refer, normalizing a (mediamatic) separation between speech and body, voice and identity. One would anticipate then that the easy accommodation of ventriloquism should translate into more realistic, that is humanistic, autonomous agents. Yet also because of its familiarity, the voice presents a particular problem for the agents that populate the discursive and material environment of ubiquitous computing. Because the artificial, synthesized voice is drawn within the ambit of media and telecommunications associated with humanism, it is already weighed down by a history and gravitas that is less formative in many other areas of ubiquitous computing. This history and familiarity place limits on the extent to which the voice can occupy multiple places and identities and is not simply a product of electronic mediation, for uncertainty also touches the non-reproduced so-called original voice. Moving from the interior to the exterior, carrying traces of the body into evanescent speech, the voice is and has always been haunted by its multiple identifications. While generally associated with the production of language, the sound of the voice also reveals the physical and, by inference, emotional state of the speaker as, for instance, being in a state of anger, nervousness, mirth, congestion, or psychosis. The temporal coincidence of hearing oneself speak at the moment of utterance, its essential autoaffectivity, has, as Derrida elaborated many years ago, grounded the voice in a presence that, although illusory, nonetheless resists interrogation. In *Echographies of Television*, Derrida extends the temporal dimension of the moment of utterance, linking it to the idea of the "event" or the experience of the event, as presented, or transformed, through "teletechnologies" (Derrida and Stiegler 2002, 3). For Derrida, the televised, transmitted, downloaded, recorded event is first "dislocated" and then "initialized" by the media apparatus, which also initializes the user (analogous perhaps to the way that computer software is initialized at startup). "The global and dominant effect of television, the telephone, the fax machine, satellites, the accelerated circulation of images, discourse, etc. is that the *here-and-now* becomes uncertain, without guarantee: anchoredness, rootedness, the *at-home* are radically contested" (ibid., 79).[5] Derrida's comments have a particular valence given that they were made at the time of their own recording (the recording of the interview between Derrida and Stiegler from which *Echographies* was produced), mimicking, in a technological milieu, the coincidence between speaking and hearing one's voice that guarantees self-presence in Western philosophy. While the proximity of the recording apparatus provides the illusion of a direct and instantaneous transfer of the speaker's intent, Derrida stresses that this proximity is also an interruption: "When the process of recording begins, I am inhibited, paralysed, arrested, I don't 'get anywhere' and I don't think, I don't speak in the way I do when I'm not in this situation. ... A modification is produced—in any case, in me ... which is at once psychological and affective" (ibid., 70–71). The event in this context becomes an artifact—something that is made or produced, that fictionalizes "actuality" and that casts a web within which producers and consumers ("subjects" and "agents") are caught. Examples of this process include "artificial synthesis (synthetic image, synthetic voice, all the prosthetic supplements that can take the place of the real actuality" but also "virtuality (the virtual image, virtual space and so virtual event)" (ibid., 6.) The process of virtuality or artifactuality is manifested by changes or determinations in temporality and dynamics: for instance, speaking on television induces an acceleration in speaking rate; the rhythm of conversation changes while the dominant presence of an accelerated speech affects the volume of those who cannot keep the pace, rendering them silent.[6]

Derrida's emphasis on rhythm is particularly important in the context of vocal transmission via "teletechnologies" and human-computer conversation. Unlike gesture or image, the voice initiates both the rhythm and the tone of ensuing dialogue and when interrupted can—as detailed by advertisements showing the interpersonal problems caused by unintentional dropped calls—terminate both a dialogue and a relationship. In using the aural metaphor, not only is Derrida referring to the recording of the interview he is currently doing with Stiegler or the speech of public intellectuals in Europe or the magnitude of the voice as a producer of artifactuality by making, and un-making, events, but he is alerting the reader to the dynamic that changes in speed produce: technological acceleration has an inverse relationship to volume, and volume must be thought at all levels, from the loudness of a particular media voice to the very possibility of other voices, other opinions being expressed, to the appropriation of acoustic space by artifactual and virtual processes, and the transformation of time into media time.[7] As we shall see, this dynamic also shapes the moment of exchange between the autonomous agent and user or client, transforming the presence, voice, and identity of both.

"Speak to Me"

As every cell-phone user knows, pauses, changes in rhythm, and interruptions can be annoying but are also an integral part of the repertoire of what researchers describe as "human conversational protocols." These include, for example, taking turns, using hand gestures and facial expressions to indicate emphasis or as listening responses, as well as changing posture, approaching or turning away, saying hello and goodbye, and so on, to signal the desire to initiate or terminate a conversation.[8] One of the main goals in ubiquitous computing research is to design interfaces—often referred to as Embodied Conversational Agents (ECAs)—defined by key researchers as "animated anthropomorphic interface agents that are able to engage a user in real-time, multimodal dialogue, using speech, gesture, gaze, posture, intonation, and other verbal and non-verbal behaviours to emulate the experience of face-to-face interaction" (Bickmore and Cassell 2005, 2). These agents simulate many of the conversational skills that humans possess in order to approach, as closely as possible, the as-yet unsurpassed richness of face-to-face conversation. A good example of an ECA in action is the prototype developed at MIT by Cassell and Bickmore known as REA (real estate agent). REA has a female figure and is capable of speech with intonation and facial and bodily gestures. Her image, projected onto a screen, interacts with users, responding to their nonverbal cues (sensed via cameras that track head and hand positions in space) and to their voice and manner of speaking (input via microphones attached to users' clothing). REA's novel features are clustered around its ability to generate gestures, to allow the speaker to interrupt its speech, and to interpret gestural and intonational cues from the user. REA's responses are produced by "an incremental natural language generation engine" that makes use of human speech and gestures, produced in real time, and activated by the user or client's vocal and gestural input, adding redundancy and thereby increasing the richness of dialogue (ibid., 5). In a typical conversation, REA would gesture to initiate the conversation and after uttering something like, "Do you like this house?" would wait for a response before continuing, using gesture to mimic the usual posture adopted while waiting for an answer. REA is designed to finish a sentence if the client interrupts using a hand gesture (picked up by the vision system) indicating that the client intends to speak or has a question, but will initiate interruption protocols (such as tilting the head) if the audio input channel recognizes overlapping vocal tracks that hit a certain audio threshold—that is, if the client speaks over REA's question or response.[9] Of course, forming the appropriate gestures, facial expressions, and intonation means nothing if the ECA's speech is unintelligible or its "voice" is unbearable.[10] As researchers note, the tonal modulations, emphases, pitch variations, and inflections that occur in speech "ground" meaning within a particular context that is locational, social, and psychological. Because synthesized speech lacks intonation and expressiveness, or affect, thus reducing its intelligibility, attempts then have been made to synthesize affect by correlating acoustic parameters with affective content: primarily through frequency (pitch) and duration. Janet Cahn describes the physiological effect of speakers' emotions on the tone of their voice:

> With the arousal of the sympathetic nervous system—as for fear, anger or joy—heart rate and blood pressure increase, the mouth becomes dry and there are occasional muscle tremors. Speech is correspondingly loud, fast and enunciated and has much higher frequency energy. With the arousal of the parasympathetic nervous system—as for boredom or sadness—heart rate and blood pressure decrease and salivation increases. Speech is slow and low-pitched and high frequency energy is weak. (Cahn 1990, 2)[11]

For modeling purposes, Cahn uses categories of acoustic indicators such as: pitch (accent shape, contour shape, pitch range, etc.), timing (exaggeration, pauses, hesitation, speech rate, and the frequency of stressed words), voice quality (breathiness, brilliance [ratio of low- to high-frequency energy]) loudness, tremor, and articulation (precision). In speech generation systems, the speaker's input is assessed and mirrored by the computer agent. If the client's speech is slow, hesitant, soft, and lacking in energy, the autonomous agent responds *as if* the client were sad (apparently sadness is the easiest emotion to recognize) or perhaps bored. There are of course side effects—for instance, raising the pitch of an ECA to mirror the voice of a client who has suddenly become very enthusiastic may result in a voice that sounds as if it is coming from a different speaker. For these reasons, researchers at MIT's Affective Computing group have opted to use "natural language" material as the sonic source for their database and analyze that material using models that automatically detect affect in speech, and employ similar analytic categories and measurement systems to those described by Cahn. Instead of using human actors, who the group believes will often produce unnatural "performed" expression, they use raw data—the "natural speech," collected from volunteers who navigate an interface guided by ECAs, who prompt the user to speak about an emotional experience, and in so doing contribute to the database. Although, as Pentland claims, it is possible to construct measures for different types of "vocal social signaling" that can be computed in real time using only a PDA, and can successfully predict a speaker's level of engagement with, and even attraction to, another speaker (Madan, Caneel, and Pentland 2004). It would seem that the unique intersection of sound, speech, and intonation that all cohere in vocal "tone" resists such measurement in all but the most regulated conversational environments. As I will discuss later, apart from practical problems such as these, projects aimed at *imitating* the effects of emotions on speech must themselves be contextualized within a system; one could call it a feedback loop—where the models used to synthesize affect, either because they don't work, or work poorly, actually produce a listener who becomes angry, sad, or bored (for instance) during the process of engagement. H. R. Ekbia gives the example of "Claire," a "virtual service representative" built in 2002 for Sprint PCS, that responds to an "unclear" user with "Hmmm ... let me get more information," and if the user continues to be unclear, transfers him or her to an operator (Ekbia 2008, 175). Other virtual service representatives are less accommodating, however, and will simply terminate the call if the user continues to give the wrong responses. Anyone who has ever been caught in the labyrinth of an automated answering service—who after half an hour going through menu items is still unable to speak to a (human) operator, or whose accent is too thick, voice too low, or speech too hesitant to be understood by the automated handler (who then abruptly ends the call)—knows that navigating such systems may easily begin happily enough only to end in frustration or anger.

In the REA interface, the clients' gestures are correlated with their words as a way of assessing their likes and dislikes, and this complements a speech-generation system that looks for emotional cues in the acoustic shape of users' speech. The gesture may be matched to a particular word through time stamping the audio and visual input data, in which case the word is "tagged," emphasis is noted, and REA follows the general command to praise or agree with a client's tastes.[12] The following exchange provides a telling example of the kind of conversation that is produced:

> Tim, the human client, says, "Show me the kitchen."
>
> REA shifts the viewpoint to show the interior of the kitchen and says, "It is a modern kitchen."
>
> Tim says, "I like the blue tiles" with a beat gesture on the word "blue."
>
> REA responds by saying, "Blue is my favorite color."
>
> Tim says, "I like the blue tiles" with a beat gesture on the word "tiles."
>
> REA responds by saying, "I love tiles." (Cassell et al. 2001, 63)

The researchers freely admit that conversing with REA is still a bit awkward, and they outline a number of improvements that would increase the bandwidth in lines of communication by adding input channels (e.g., measuring head movements and eye gazing to estimate the direction of the client's face) and increasing the number of conversational protocols that REA can respond to. It is fairly easy to imagine far more sophisticated and streamlined conversations in the future, dependent, of course, on technological advancements in the field. However, the clunkiness of REA's responses points to more than a technical inadequacy, revealing something of a black hole at the heart of the very enterprise to simulate authentic response from computer agents. As Cassell and Bickmore note, users found REA to be "unfriendly and cold" and sometimes preferred conversing with it by telephone. Apparently REA could not sustain the suspension of disbelief required to elicit trust and engagement in the user—a task that requires an almost perfect simulation of human behavior. According to the researchers, the issue is not, however, that users failed to summon the necessary credulity but rather that REA *failed to convince* users who would otherwise have "automatically" responded to an ECA as if it were human. "Our experience has been that belief in a computer agent's acting like a person is automatic from the first moment of an interaction, and it is this belief which must be 'suspended' by the user, when the agent fails to meet their expectations by behaving inappropriately" (Bickmore and Picard 2003). Further, researchers note that even though participants in experiments declared that they would never follow conversational protocols when interacting with a computer, at the same time, during the interaction they apparently did. Obviously contradictions abound in this particular environment, and during this moment of human–computer exchange, users need the ECA to continually maintain an illusion regarding its nature that is sufficiently powerful enough to engage them in a form of interaction that they disavow the minute the conversation comes to an end. While users didn't attribute humanness to REA ("nobody was going to leave thinking that this was a new living species, or a new kind of human"), at the same time this particular ECA, being cast as "cold," apparently wasn't human enough (Cassell 2000, 7).

Are You There?

Failing to maintain the illusion of humanness has implications for HCI as a whole, for one of the research objectives has been to create interfaces that would lead the user to have more trust in technology. Trust, the researchers emphasize, is essential for sociality—without it, conversation is thwarted: "Interactions between two people who do not trust one another are difficult to sustain: they display less verbal fluency, are filled with pregnant pauses, with incoherent sounds, with dropped words" (Cassell 2000, 1). Dialogue becomes "telegraphic and formal, approximating a command language," people speak more carefully and less naturally, their words becoming both measured and mannered (Bickmore and Cassell 2005, 11).

If lack of trust produces the kind of awkward dialogue that occurred in the example above between Tim and REA, it also reproduces the kinds of conversations associated with both early telephony and contemporary wireless. Characterized by frequent pauses, unintended interruptions, and the reoccurring questions "Can you hear me?" and "Are your there?" the degraded "presence" of (tele)presence has been, to some extent, addressed by two related, recursive operations—first, the increasing sophistication of transmission technologies, and second, the adaptive techniques developed by auditors. The latter often include ways of dealing with the tone of voice, since the temporal and sonic peculiarities that constitute tone (rhythm, modulation, volume, accent, and so on) do not transmit very well. In fact, tone transmits so poorly that there are only certain modulations that one would attempt when speaking, for instance, on the telephone. As a result, hesitation, offhand comments, various forms of sarcasm, and anything that requires a low voice or various pauses has been filtered from phone conversation. The actual discontinuity in speech and communication that telephony first imposed has been amortized through decades of improvement in the quality of vocal transmission, culminating in the heyday of landlines during which long and often highly emotional conversations seemed to dematerialize the apparatus itself. This technological and historical continuity seems to embody the overarching conceit of technocultural determinism—that with sufficient progress (or enough bandwidth), all human attributes can be transmitted (or downloaded).

It comes as no surprise then that the quest for noiseless telecommunications would develop into the larger project to overcome the "noise" of distrust, disinformation, and disinterest that new substitutes for face-to-face interaction, such as ECAs, might generate. Yet this larger project carries with it the remnants of an already filtered, adapted, tailored, somewhat muted telephonic voice. The delayed, diffracted, pause-filled conversations of early telephony, the short, barely audible, and abruptly terminated voice of the wireless caller, the unpleasant and often unintelligible voice of the ATM, and the not-human-enough ECA are bound together by an acoustic content, or lack thereof, which, in the context and moment of ubiquitous computing has an existential correlate in the degree to which the ECA appears to be "embodied" or the client feels himself or herself to be disembodied, to be "not-at-home," as Derrida has said of teletechnologies, in the conversation. If there is a gap in the flow of dialogue, if the ECA seems to lack the warmth associated with human presence, if there is a degree of distrust generated by the inadequacy of the simulation, the entry point of these separations has its origins outside of this particular instance in the genesis of REA's speech and in the media-specific context of its conversations. Thinking about the circumstances of such narrations, the mere fact of talking to a computer, talking into a microphone, or talking to a telephone receiver may subtly change the voice, contaminating it with the unnaturalness that designers are trying to avoid. Of course, as Derrida and many other writers have pointed out, there is no "natural" voice, and the notion that the voice provides an origin-grounding speech in unmediated presence is a convenient myth. However, as I have discussed elsewhere, a belief in the "original," unmediated *sound,* a sound that could be transferred ontologically intact through (analog) recording, persists as an operating principle in the common understanding of the digital recording.

In the conversation between Derrida and Stiegler mentioned previously, Stiegler argues that digital recordings, for instance, open up a new relation to the future, "precisely in that they make it possible to capture *exactly* the grain of the voice, the body, and by the same token, transform this body and its psyche" (Derrida and Stiegler 2002, 102). Here Stiegler is referring to Barthes's notion of the grain of the voice, exemplified by the singer Charles Panzéra (whose phonographic recordings had, according to Barthes, an "electronic purity") and the "reality effect" that the reproduction of grain (through the analog recording or photograph) triggers in the viewer. Stiegler expands the notion of the reality effect to what he calls the "authentification effect," which relies upon the accumulation of exact reproductions, producing, over time, a sense of "presence," creating the conditions for the "evolution" of our perceptual beliefs, and indeed, a new form of intelligibility according to a logic that is unique to the digital (ibid., 149).[13] As he writes, the "analogico-digital image," because of the possibilities for manipulation that it offers, also enables

> new forms of "objective analysis" and of "subjective synthesis" of the visible—and to the emergence, by the same token, of another kind of belief and disbelief with respect to what is shown and what happens. ... By discretizing the continuous, digitization allows us to submit the *this was* to a *decomposing* analysis. (ibid., 152, 157; emphasis added)

If this new intelligibility—an intelligibility that might plausibly be associated with posthumanism—depends on a new understanding of the image that is no longer influenced by the belief in an authentic original, but rather can see the manipulations and the mediations that have occurred, then we might wonder how such "cognizance" incorporates the aural. With the digital recording supposedly preserving the sonic properties of the original affective expression, the transmission of attitudinal or emotional states would seem to require no more than a methodology for transcription and translation. Yet the critiques of the ontologically intact recording offered by sound theorists decades ago are salient in this context. First, the recording is already abstracted—removed from its real-world context, the conversational mode of its utterance and the emotional state it represents. This is as true of the automated voice as it was of the recorded voice of analog media. As Pantic and Rothkranz note, "In many instances strong assumptions are made to make the problem of automating vocal expression more tractable." For instance, the recordings assume a fixed listening position, a closely placed microphone, a recording environment that is noise-free; while "the recorded sentences are short, delimited by pauses, and carefully pronounced to express the required affective state" (Pantic and Rothkranz 2001, 470). Second, as mentioned, the presence of the recording apparatus affects the speaker in ways that bear directly on the interpretation of affect as a composite of sonic features. Derrida points to this unintended staging in his discussion of teletechnologies with Stiegler, describing it in terms that emphasize the aural nature of the recording's predisposition:

> With all these machines and all these prostheses watching, surrounding, seducing us, the quote "natural" conditions of expression, discussion, reflection, deliberation are to a large extent breached, falsified, warped. One's first impulse would therefore be to at least try to reconstitute the conditions in which one would be able to say what one wants to say at the rhythm at which and in the conditions in which one wants to say it. And has a right to say it. (Derrida and Stiegler 2002, 32)

It might be tempting to see in Derrida's critique of teletechnologies the desire to reconstitute an original, unmediated voice. However, Derrida emphasizes that technology has always been present in and with the voice: "These machines have always been there … even during so-called live conversation" (ibid., 108). In the same way that technologies have always infiltrated the voice, transforming it into an "instrument," there has also never been a "this was" (per Stiegler) that pertains to sound and the voice. Even with the recording endowing the voice with a degree of endurance and the metaphysics of the voice providing the illusion of presence and identity, the "this was" vanishes in the ephemerality of sound. Immaterial and temporal, described in terms that always defer to the material object ("the" sound of a plane, the door, the glass breaking), sound struggles with "this" (for sound is always multiple) and relatedly with the "is" or "was" that endows the object with persistence, even within the illusory reality of the image. Because there has never been a "this was" for sound, the aural correlate of the digital image—what Stiegler might call the "analogico-digital recording"—fails to call into question the authenticity of the original sound because that of which it might be an indexical trace is not itself an object. The audio recording doesn't inspire the same kind of fear in the listener as the image—that it is both true (it was) and false (it may have been manipulated) at the same time. Rather, the recording inspires a different kind of fear: the fear of the duplicity of the voice as such. For just as the voice reveals affect—the supposed "truth" that lies beneath the speaker's speech that transmits, far more than words alone, the speaker's intent—so it may also be affected. Distinguishing between affect and affectation, "true" emotion and staged performance, is a skill, knowledge, or intelligence that is practiced during the conversation, in real time, and requires a level of concentration and engagement that exceeds the storage, prediction, identification, and production capacities of the average autonomous agent. In contrast to the unique intelligence, the "more knowing belief" (ibid., 152) that Stiegler attributes to digital technology as it reveals that, for instance "the analog image is always-already discrete" or "animation is always reanimation" (ibid., 155–156), the awareness of the listener/conversant is not so much concerned with the origins, constitution, or genealogy of the voice it hears. In other words, this *listening* is not concerned with a potential for speculation, analysis, or objective ("theoretical and scientific") knowledge, which is an intelligence or consciousness that belongs to the visible and presupposes a metaphysics of the object. In the context of posthumanism, and what it does or does not offer, we might wonder how this "intelligence"—a product of the evolving perceptual capabilities associated with our partnerships with intelligent machines—operates when there is no object, no analysis, only the enigma of tone heard in the ephemerality of the moment. How does it codify, compute, or translate the tightening of the throat, the dryness of the mouth, the muted, hoarse words, or the stunned silence that is the experience of the other, the unexpected and the unanticipated, the experience that exceeds the conversation and yet is necessary for the conversation to proceed? The knowledge of the origins of the voice—whether the voice issues from a loudspeaker, a screen, or a cell phone, whether one knows (Stiegler's "new form of knowledge")[14] it has been manipulated or not—is irrelevant in this context. For the conversation already presupposes an engagement and commitment, and it is the veracity of this engagement, its intimate relationship to and effect on the human speaker, that is at issue in the dialogue with, and debates about, autonomous agents. The stakes are high in these moments, for questioning the being of one's conversant is also ineluctably questioning the being of oneself.

Hating Microphones

Regardless of the intention to imitate natural speech, ECAs must also imitate the models of transmitted speech that listeners are accustomed to. Gesture, intonation, facial expression, and body posture are mapped back onto the most familiar models we have thus far of amplified voices: television news anchors. These are also embodied (at least from the waist up) and localized in a particular geographical place—as indicated by the (possibly simulated) background of the television set. Context is everything, and the many eras and modes of media that form a backdrop to REA's appearance have already embedded certain modes of viewing figures on screens or listening to voices emitted from speakers. Although rarely dealt with in the literature, the audio apparatus plays a decisive role in the intelligibility of the ECA's speech and its receptiveness to the speech of the client or user. Too little amplification, or incorrect filtering, and the ECA's utterances cannot be heard—end of conversation. Because an acoustically neutral background is dominant in the majority of media voices, it is also *anticipated* as part of the listening and conversational exchange. Indeed, an overly noisy environment that interrupts speech is seen as an obstacle to be overcome by adopting certain audio techniques. The presence of unacknowledged production protocols thus adds another track, or dimension, to this form of human-computer articulation, smoothing over both the fluctuations in the external environment and the clunky strangeness of the ECA in the conversational exchange. This is not to claim an absolute equivalence between one form of media and another or to deny the specificity of ubiquitous computing but, rather, to suggest a genealogy of the ECA and its corollary, the listener, that acknowledges the production of both within an environment already populated with loudspeakers, amplifiers, transmitters, receivers, screens, and all the various components that go into creating the media ecologies we currently inhabit. Telepresent and human-computer conversation are dependent on and formed by these prior cultural/technological infrastructures, a fact that REA acknowledges in its second conversation with Tim:

REA That microphone is terrible. I hate using those things.
REA Sorry about my voice. This is some engineer's idea of natural sounding.
REA Are you one of our sponsors?
If the user answers "yes," REA asks if he or she was at the last sponsor meeting and then refers to her voice once again.
REA I got so exhausted at the last sponsor meeting. I think I was starting to lose my voice by the end. (Bickmore and Cassell 2005, 15)

REA's confession is aimed at preparing the ground for the more practical discussion of where Tim would like to live and what kind of house he has in mind. By having REA disclose some of the more intimate details about its "life," the designers hope to elicit a similar response in the user and in the process establish some sort of bond. While the self-disclosure in this case purportedly concerns the limiting features of the technology, it also refers to REA's constitution—its past, its genesis, what might be called its ancestry. The user learns that REA does not "own" its voice, that its voice is rather the manifestation of a human idea—an engineer's idea of what a "natural" voice would sound like. REA's voice is a poor simulation, the poverty of which REA acknowledges, indicating to the user that it is self-aware enough to know that it is a computer interface and self-conscious enough to feel embarrassed about its inadequate construction. It might be tempting to see in such awareness the signs of an autonomous entity coming to life and feeling the twinges of an identity crisis. But what vocal identity does REA really have? While the dialogue between REA and Tim seems to be a unique interaction, tailored to the individual user, REA's speech is in fact generated from a database that has been built from personal narratives related by other users to other ECAs.[15] This construction in itself has implosive tendencies since, as much of the literature acknowledges, it is almost impossible to design ECAs that will seem friendly and warm, that will respond appropriately and will be "affect-sensitive" without using so-called reality-mining systems, which are notoriously prone to dissimulation.

Indeed, the problem of dissimulation cascades through the systems, interfaces, and devices used in those areas of ubiquitous computing that identify, register, and simulate affect. Tracking bodily indicators such as heart rate, galvanic skin response (sweat), breathing rate, gait, and facial expressions in order to identify and measure a person's emotional state assumes at the outset a calculable correlation between changes in emotional states—what principle researcher Rosalind Picard refers to as "sentic modulation"—and certain physiological patterns.[16] Not only do pattern-recognition technologies embedded in specially designed devices that are wireless, wearable, and spatially deployed inherit the often crude correlations between physiology and psychology, they also rely upon what are essentially artificial perceptual systems—remote sensing, audio and vision systems, and facial and vocal analysis—to collect data, and artificial intelligence systems to analyze and represent them. At both ends of the perception/knowledge loop, then, there are the mediating influences of computing (with its quantitative interpretative methods and modeling limitations), based as they are on the reduction of mind and matter to a computational paradigm. Hubert Dreyfus has pointed out that the amount of information collected by wearable devices, pale in comparison to that stored already in the body, and therefore "all the zillions of facts that are relevant to the body and that the body knows ... can't be accessed" (Dreyfus 2001, 18). In addition these systems cannot access the repositories of affective expression that are constantly being repressed or manufactured in social situations—a problem that MIT researchers working in the field of affective computing and "e-rationality" recognize. Picard notes the difficulty in finding the "real emotions." Not only do emotions vacillate, but joy and anger "can have different interpretations across individuals within the same culture." Finding the true emotion, however, is almost impossible when the subject knows he or she is being monitored or is part of an experiment. With a new iteration of Heisenberg's uncertainty principle, Picard repeats the dictum "measurement of ground truth disturbs the state of that truth" (Picard, Vyzas, and Healey 2001, 1178).

If the quest for true emotion risks encouraging the *simulation* of emotions rather than their direct experience, affective computing itself as a field of research risks interposing the same ambivalent dynamics of belief that the attribution of emotions to machines elicits. For if it is enough that computers mimic rather than experience emotions, perhaps this holds true for humans as well? What would be the consequences of such widespread simulation? What kind of subjectivity would develop as users—that is, anyone who travels, banks, phones, uses the Internet, or occupies public spaces—become increasingly cognizant of their electronic doppelgangers? Could the knowledge that every emotional indicator is being monitored induce a kind of social and biological autism? Or would the feedback loops generated between the device and its wearer instead create a sophisticated repertoire of postures designed solely for the camera, while deskilling, through under-use, spontaneous emotional expression when the camera isn't there? Picard raises the issue of autism in relation to online communication:

> In a sense, everyone is autistic online: today's systems limit your ability to see facial expressions, hear tone of voice, and sense those non-verbal gestures and behaviors that might otherwise help you disambiguate a hastily sent non-angry message from a genuinely angry one. To the extent that people spend more time communicating with each other without sufficient affect channels, they may actually be reducing some of their emotional skills—a kind of "use it or lose it" opportunity cost. (Picard and Klein 2002, 157)

Autism figures within the literature as both a (technological) impediment to full communication and a (human) disability that technology might aid.

In many of these projects, the underlying premise is that all communications technologies and some people (e.g., those diagnosed with autism) lack what the researches refer to as "mind-reading" functions—that is, the ability to read the mind through the facial expressions and vocal tone that accompany most human-to-human exchange. In this respect, humans and machines meet on the common ground of a disability that can be addressed technologically only by disassembling the circuit between emotion, expression, representation, and language. This occurs via the modeling systems used, which examine acoustic parameters for indicators of affect and then target the affect in speech rather than the meaning of the words spoken. Pentland describes the working methodology as "disambiguating the affect of the speaker without knowledge of the textual component of the linguistic message," implying that meaning interferes with the transmission of emotion and somehow needs to be put aside or ignored.[17]

Registered through acoustic parameters (such as volume and rhythm) or conversational patterns (such as the dominance of one speaker or the frequency of interruptions), affective content thus becomes an object of therapeutic interrogation and intervention. The tone of the voice, the particular vocality that represents and carries affect, enters this problematic as irreducibly "ambiguated" in that it always carries both meaning and expression, and this ambiguity is exponentially multiplied when the speaker knows in advance that he or she may not be speaking to another human and that his or her speech may be monitored by machines that might approach a crude approximation of hearing but certainly do not listen. Dissimulation appears in this context as an even more tightly bound contradiction. Installed in the workplace, for instance, modeling affect in speech may produce a "performed" autism that is itself a reaction or adaptation to the various prostheses devised to counter the technological autism inherent in machines that have no senses. If any of these technologies are to have a general rather than individual application, the database must be enormous. Yet the garnering of affective instances through constant monitoring means that the dissimulation provoked by surveillance will itself block any access to the truth of the subjects' text, speech, or what ultimately becomes their performance.

Concerns over privacy issues have led researchers such as Picard to call for voluntary participation in contributing by recording their conversations using a wearable device. Overseeing and supervising one's conversations in this way may be impossible because, in contrast to text-based correspondence, conversational speech is volatile and requires far more self-control if one wants to keep certain views, opinions, and feelings in check. But more importantly, it begs the question of how aware we are and how much control we have over our vocal tone. We should remember that just a voice might express affect; it can equally be "affected" marking the voice as an instrument. Indeed, if tone is registered by acoustic parameters that are independent of or rather run as a lyrical or musical correlate to what is spoken, they cease to be strictly acoustic but are better understood through the musical meaning of tone and the performative meaning of intonation. Tone is both more than the sound of speech and less than the meaning of the words spoken. Like physical gesture, it shares with music and sound an ambiguous status, an ill-defined meaning. Tone can slip into the sound of speech, just as the sound of speech can disassemble into babble or accelerate into fury, rage, or an almost silent, speechless hiss.

If both language and tone must be carefully monitored by the speaker at the same time he or she is attempting to engage in spontaneous, trust-building behavior, "[then] whether consciously or not speakers will limit acoustic markers such as volume and pitch (which might indicate anger or frustration), pacing (which might reveal nervousness if too fast, or boredom and fatigue if too slow), modulation, and inflection (which might indicate unintended sarcasm)" (Eagle and Pentland 2003a). In other words, the conversations garnered for the database become a series of performances where both semantic and expressive content of exchanges is "scripted" prior to their transcription via text-to-speech programs and their eventual revocalization as responses by ECAs. When these ECAs then converse with participants in order to garner affective content to be used by other ECAs, the degree of performativity is multiplied. We need to ask then, if speech on both sides of the human-computer interface is to varying degrees "affected," to what extent is authentic interaction possible? Doesn't the simulation of (the possibility) of dialogue in one context lead to the simulation of, say, emotions such as trust in another? Doesn't the ventriloquism of REA rebound in an endless series of performances—each becoming the origin myth, the "true" voice, or "true" relationship of the next? In a process similar to the acoustic degeneration associated with the reproduction of analog recordings, the contamination of signs of affect, signs of trust, and signs of authentic interaction builds as each snippet of conversation, each response, is circulated through the various databases that form the backbone of the ECA's speech. In the same way the sound is lost in analog reproduction, it seems that in this case, while actual sound quality might be preserved, the link between affect and its expression is compromised.

Loving Too Much

Exhibiting the self-awareness usually associated with human beings, REA both disowns its voice and claims an equality with its human interlocutor because presumably only a human would "feel" embarrassed about its voice, would apologize for its "unnatural" synthetic sound, and would have the sensibility to know the difference. Perhaps because of the unnatural sound emitted from the speakers and then synchronized with the digital figure on the screen, REA does not like having its voice amplified. In fact, it "hates" microphones, revealing an attitude, a preference, perhaps even an emotion that furthers the process of human-machine bonding. But as mentioned, hating microphones also references REA's genesis from the decades of vocal production, amplification, and transmission that are generally associated with media. Not being media allows REA to claim a degree of uniqueness, perhaps the uniqueness associated with being human, or at least, being alive. Not being media shifts the discourse from the veracity of the simulation (with its attendant implication of "unreality" and potential deceit), to the awkwardness of REA's apology and confession, as if it were something to be taken seriously, as if REA were a thing to be taken "at its word." During this moment, Tim might be tempted to respond humanely, to disavow the obvious, to invest a little in the fantasy of REA's "life," and in so doing, might activate the very processes that autonomous agents depend on: anthropomorphism and the discursive shift from AI to A-life, the shift that frees the agent from the solipsistic bonds of human engineering and human thought, the shift that allows the agent to be its own idea. Manifesting the physical signs of life—movement, speech, and so on—having a humanoid form, and speaking of "love," REA is already primed to elicit an anthropomorphic reaction in the user. As Hayles notes, "Mystifying the computer's actual operations, anthropomorphic projection creates a cultural imaginary in which digital subjects are understood as autonomous creatures imbued with human-like motives, goals and strategies" (Hayles 2005, 5). Anthropomorphism isn't limited to agents having some kind of humanoid figure but extends to the belief in technological agency that may or may not be "embodied." The self-awareness that REA exhibits in apologizing for its voice is in fact the "awareness" of a sophisticated retrieval system and could be compared to the awareness that Stiegler attributes to digital technology in its ability to expose prior mediations. But the synthesis leading to a new understanding that digital technology triggers is imminent to the technology only insofar as it has been equipped with a database of signature events—events that resound without context, outside of the moment of their issuance, devoid of the "acoustic universe," as Adriana Cavarero writes, from which language itself arises: "The vocalic's base is thus essentially musical in the sense that it conforms to the contextual sonority of the world, to its noises: and it participates in it as well" (Cavarero 2005, 148). The database is, if you like, always already dislocated from itself and only ever able to issue a particular interpretation, context, and worldview that simulates, but at the same time extinguishes, experience. This is not a self-reflexivity or self-awareness as much as it is a form of solipsism, operating at a systemic level and placing limits on the goal of the conversation as an appropriate metaphor for human-computer design.

For Picard and Klein, the practice of anthropomorphism can in fact destroy human-computer interaction: "A few human-like features poorly implemented can be much worse than no human-like features at all. This phenomenon is already well known in computer graphics, where interaction with a supposedly realistic humanoid character still leaves the human viewer with a more eerie and disturbing impression than a corresponding interaction with an intentionally non-realistic character" (Picard and Klein 2002, 10). Such feelings of eeriness are compounded by the artificial voice, the voice with its multiple identities. If the media alters the experience of the event, turning it into a "media event," we might wonder whether this "initializing" process operates in the context of artificial voices.

Mladen Dolar, in his insightful book *The Voice and Nothing More*, writes, "The voice is like a fingerprint, instantly recognizable and identifiable" (Dolar 2006, 22). Although not contributing directly to meaning, this uniqueness of the voice—its particular tone, timbre, inflection, accent, pacing, and register—is necessary for both the simple transmission of meaning and the identity of the speaker to be established. When the voice is synthesized, both meaning and identity are uprooted, and in the place of stability (however illusionary) there is only the orphan voice of an automaton. As Dolar writes: "The impersonal voice, the mechanically produced voice always has a touch of the uncanny … [for it] reproduces the pure norm without any side effects; therefore it seems that it actually subverts the norm by giving it raw" (ibid.). The irony of the artificial voice—the kind of voice that might belong to an autonomous agent—is that it can never be good enough, even if it is perfect: "The voice without side-effects ceases to be a 'normal' voice, it is deprived of the human touch that the voice adds to the arid machinery of the signifier, threatening that humanity itself will merge with the mechanical iterability, and thus lose its footing" (ibid.).

The experience of uncanniness that the inadequate simulation produces, such as REA's overexuberant response to the mention of floor tiles, far exceeds a simple disappointment in the sophistication of the technology. No amount of programming will diminish the feeling of eeriness that hovers over the entire exchange with REA, for the emotional content of this ECA's response—even if not awkward, lagging, or glitchy—is both excessive and out of place. It is not so much that REA is "overacting" or its eagerness to agree with the user destroys its credibility as an independent, autonomous real estate agent (rather than a program), but its response begs the very question that the user wishes to be deferred in engaging, presumably sincerely, with an ECA, that is, the question of the ECA's autonomy and subjectivity, introduced by its use of "love." Without an authentic conversant, the user's speech resembles a soliloquy, and the experience itself registers the force of an encounter with a nobody. At the instant of this recognition, the user, having held the floor, now finds him- or herself disqualified and, incidentally, speaking in a strange manner to a figure on a screen. In this moment, the user might experience the kind of paralysis that Derrida articulated in reference to the recording, s/he might be stopped in his or her tracks, perplexed by a doubt that comes from nowhere and arises unexpectedly. Indeed, the user could be suffering from a form of buyer's remorse, a moment of hesitation that interrupts the celebration of technology's brilliance by wondering where to put it, how to place it, how to keep it at a distance, how to love and hate it at the same time.

Moving between automated answering service, traditional screen-based animation, science-fiction character, and the interactive audio/visual display viewed on screens positioned in the lab, the showroom, or in this case, perhaps in the local real estate office, the shifting levels of interactivity manifested by ECAs also define their status as media or machine, simulation or robot. The formation of the ECA's identity is influenced by popular cultural representations of robots and cyborgs as entities that share human traits, such as the ability to feel. While more the stuff of fantasy than actuality, the figure of the android that is too humane to be human (*Alien Resurrection*) or the replicant that has to be retired before it develops emotions (*Bladerunner*) casts a shadow of expectation over computer agents (a.k.a. interfaces) such as REA. It may be that in loving too much and too quickly, the ECA threatens the quality of "real" emotions, which, in consumer society, may be the last bastion of pricelessness: those "human" qualities that cannot be bought and sold. Yet if popular science fiction is any gauge of public sentiment, then when computers love too much, all technophilic fantasies implode. An excellent example here is Steven Spielberg's film *Artificial Intelligence,* influenced in part by the ideas of AI pioneer Marvin Minsky, who, according to Ekbia, was involved as an advisor (Ekbia 2008, 320). Based on the fairy tale *The Adventures of Pinocchio*, the narrative is centered on the absolute determination of a robot child (David) to win the love of his human "mother" (Monique) and thereby become human.[18] David, the first of these robots, was programmed to love; however, he was not given the tools to contain or regulate his love. David's desire could only be driven to excess, an excess that matches both human overconsumption and the irreconcilable quest to find an other—unique and autonomous entity—that is both controllable and uncontrollable, that will love but only to a point. This quest drives the film's narrative, which begins in (human) tragedy and ends in absurdity.[19]

While matching David in neither sophistication nor desire, REA provides an allegory through which many of the fantasies surrounding human-computer interaction become manifest. As Kathleen Woodward reflects, the attribution of emotions to computers, robots, and cyborgs is more fantasy than fact, and its fantastic qualities "serve as a bridge, an intangible but very real prosthesis, one that helps us connect ourselves to the world we have been inventing."[20] The key to believing that a "bot" has emotions is intermingled with the same kind of suspension of disbelief that science fiction films demand of the spectator and autonomous agents require of their conversants. Yet just as this suspension produces a form of "future fact" (Woodward 2004, 191), it also just as awkwardly generates an approach toward technology and posthumanism that is productively inconsistent and contradictory. Like "future fact," the oxymoronic nature of "productive inconsistency" calls attention to the *process* of distinguishing (and not distinguishing) human and machine via the voice, a process that enables multiple attitudes toward technology to happily coexist, while, on a purely practical level, allowing consumers to find a place for all the screens that negotiate their networks and the gadgets that fill their pockets. This simultaneous celebration and distrust also transfers the reality of technological objects—the fact that they are always in the process of becoming cheaper, faster, smaller, and more productive—to the reality of the consumer. The continuous upgrading and replacement of small, portable, and affordable gadgets, and the increasing sophistication of the interfaces they contain, displaces the image of gargantuan machines dwarfing automata-like workers associated with previous decades and in the process reformulates the equivalence between machine obsolescence and human death. The potential for unlimited replaceability or regeneration is one of the conditions associated with the movement from humanism to posthumanism, marking an attitudinal change from separating to incorporating technological devices while still distinguishing human from technological existence.[21]

Yet there is a degree of mental gymnastics involved with these incorporations and distinctions that, in suspending the self, also paralyzes the thinking of the other, whatever that other might be. Rodney Brooks's proclamation regarding the possibilities of bioengineering is salient in this regard: "There is no need to worry about mere robots taking over from us. *We will be taking over from ourselves* with manipulable body plans and capabilities easily able to match that of any robot."[22] Sarah Kember cautions against the twists and turns produced by the generation and degeneration of metaphors, asking, "If the human-machine metaphor is literalized and subsequently de-literalized then what remains of it? ... If we are not proposing to look at humans as machines, machines as humans then what does the conjunction of human and machine currently mean—what is its purpose?" (Kember 2003, 197). For Kember, this conjunction functions as a bridge—not to posthumanism, but to a rearticulated humanism, signifying a desire for a "post-liberal humanism" without the critiques of post-structuralism or the cynicism of post-modernism. As such, it signifies an innocent, fantastic embrace of technology by the individual, now able to converse with a nonthreatening "agent," who, as the confessions of one autonomous agent reveal, will always yield the floor. But this conversation and this "takeover from ourselves" occur only through the pauses and interruptions, the uncertainty and hesitance that rebound on any simple partnership with intelligent machines that we may, as a culture, choose to fantasize. Like the voice, this uncertainty goes to the core of technics, which, as Derrida points out, "does not belong by definition, by virtue of its situation, to the field of what it makes possible."[23] If, in the (omni-)presence of ubiquitous computing and the transformative time of posthumanism, the thinking of technology, like the thinking of sound and the voice, threatens to disappear in the moment and in the noise of the momentous, the task must be to follow the echoes of this noise, the volume of its displacements, the awkward silence it elicits, and the voices it catches—in short, to attend to the suspensions that modulate the enigmatic, ambivalent hyphen between human and post.

5 Resonance, Anechoica, and Noisy Speech

Today, if something like a "philosophy of nature" is possible in a new way, it is a philosophy of confines. We are at the confines of the multidirectional, plurilocal, reticulated, spacious space in which we take place. We do not occupy the originary point of the perspective, or the overhanging point of an axonometry, but we touch our limits on all sides, our gaze touches its limits on all sides.

—Jean-Luc Nancy (1997, 40)

Badiou's ontology inserts a "decision" in place of the cosmos, or pure multiplicity, that might see power drift into the ether of virtuality.[1] The decision is ideological, and ideology, as Žižek explains, "exploits the minimal distance between a simple collection of elements and the different sets that form part of this collection" (Hallward 2003, 90). The organizational aspect of ideology results in a "structural repression of that part which … having no discernible members of its own, is effectively 'void' in the situation" (ibid., 89). In elaborating the role of ideology in Badiou's ontology, Hallward uses the example of music and language—the excess of possible combinations of notes and letters, producing words and music, over the relatively small number of notes and letters themselves.[2] As we have discussed, the concept of tone contributes to music insofar as it denotes a relationship between notes. The tone is foundational for music and for the entity of the note, but it would be a mistake to assume that a particular tone is wedded to, for instance, a particular pitch or frequency. It is, in the Badiouian nomenclature, an "*ur* element." In other words, while it is fundamental to the set of notes that constitute a musical sequence, it is not in itself "musical." Like the empty set, "tone" in itself, is no "thing," but, like number, is only purely relational.

This is the essence of tone—its pure signification occurs in the tone of the voice, in the unaccountable in music, the moment of recognition that flutters across a room following a cough or a sigh, in the silence before applause, and in the strangeness of poetic speech. Tone masks unaccountability—in both senses—that which can't be counted, and that which is not required to answer for its actions. If not musical, then could it be that "tone" is strictly organizational, and therefore hides, or enables, a structuring that is ideological? We have seen that in music, the concept of the tone was used to demarcate "rational" intervals on the one hand, and a representation of divine hearing (and therefore what counts as properly musical) on the other. While strictly intramusical, tonality nonetheless provided a rationale for the dominance of the "One": the origin and ground of all human and cosmic existence, and for an ontology that would lend itself to quantification based on infinite divisibility and unchanging identity. As Pythagoras's experience in the forge demonstrated, the material, sonic production of consonance would forever be ignored in the abstract system of tonality that eventuated. The relationship between this music and political theology is, as Agamben has shown, so intertwined with representations of power and divinity that it is almost impossible to disentangle.

If tone in music aided the evolution of political theology into economic theology, the tone of the voice in language has provided a hinge upon and through which politics and power are realized. Returning again to Michaux's "sans"—the broken instrument attached to a broken body that produces broken sound—it is interesting that Michaux's text is interpreted as representing the sound of the "cry" that the instrument makes via reference to a "humanized" void—"the other." The sound is extracted from any association with music and instead is likened to a voice caught in the throat, a sound repressed, interpreted as an "intermediary," an ambassador perhaps to "the exiles of derelicts of the world. ... The community of the underprivileged, imprisoned and desolate."[3] This cry demarcates access to property, space, and self-determination based on a more or less fixed power structure. On the periphery, the dispossessed articulate or at least "sound out" the border between those who possess and those who have had their property and their rights taken from them. We imagine people living on the outskirts of town in dire poverty, wandering, homeless, a diaspora (*dia* meaning "through") pouring through national boundaries in times of war.[4] This representation captures the ambivalence of the "sans" cry—because the image of pouring out is both one of generosity and plenitude, an overabundance where the cup runneth over, and the emptying out of an occupied land. The angels' voices *ring* out, capturing space; the dispossessed and broken cry *empties* out, half-heartedly, surrendering space. The capturing of space by the voice occurs through a choir of imaginary agents, devoid of physicality, whose sole raison d'être is to multiply and extend the glory of God—a glory that is essentially meaningless, as it can neither be increased nor decreased. The surrender of space occurs through the broken voice of a broken body, emitting a "throttled cry." Perhaps, the sound caught in the throat is unable to escape because of self-censure; perhaps it has been forcibly blocked. Regardless, this is speech that cannot be properly moved from the interior to the exterior, that cannot properly take its place outside of the body but, rather, reverberates or suffocates within. In not escaping the body in an orderly and one could say "administered" manner, the voice does not occupy space. It does not reverberate within the atmosphere. It does not reflect off any surfaces other than the broken internal apparatus that produced it. It does not resound, and it has no response. The ambiguous "broken" fulfills two functions here: it designates the outskirts of property—the periphery where only the dispossessed exist—and the fringes of both music and language. In so doing, it exposes the fact that so-called toneless music, in stripping away structure, leaves only sounds that in Badiou's terms occupy and indicate the "edge" of the void.[5]

For Badiou, while the void is beyond presentation because it can never belong to a situation, it can still be named. As Hallward writes, "Ontology demonstrates that, considered in its be-ing, and whatever the situation, the name of the void is the name normally deprived of all meaning and resonance, a name for anonymous namelessness as such" (Hallward 2003, 101). The mark of the void (Badiou has chosen the Scandinavian letter Ø) unites being and nothingness, and founds an ontology "without any reference to the transcendent, without any reliance on figural approximation" (ibid., 102). Only on this basis can the finite be absorbed within the infinite and can ontology escape what Badiou has termed "poeticization." However, it could be argued that poetic speech names, *through sound*, the "edge" of the void, and this is especially the case with sound poetry, the performative voice, oratory, and the presence of onomatopoeia, rhyming slang, alliteration, assonance, repetition, and so on. Like the throttled cry of Michaux's broken instrument, poetic speech de-structures language, revealing the "no-thing" of speech.

Plato attempted to circumvent the abyss between speech and the world in *Cratylus*, where he aligned the sound of the names of things with divine intent such that speech automatically became a form of acclamation.[6] This alignment is carried through to the middle ages by Paracelsus, for whom "all things bear a sign that manifests and reveals their invisible qualities" and whose "art of the signature" aligns language, the name, and the sign of the thing within the act of naming itself (Agamben 2011, 33–35). However, without this Platonic caveat, it remains the case that such speech walks a very thin line between sense and nonsense, propriety and impropriety.

Poetics is the entry of tone into language, both as a "musicalization" and a literal call to the set or element that cannot be named, which is the non-set of sound. It presents "tone" as a meaning, carried through the voice, that cannot be contained within language, and sound as a material/spatial phenomenon, as that which cannot be contained within the immaterial non-place of calculation.

As Nancy points out, however, such speech poses the dilemma of structurelessness and poses the challenge of how to configure a space without inserting "a totalitarian 'truth'—a 'truth' that structures space and sense politically"—while still making sense. Most importantly, Nancy asks: "What outline would retain the unexpectedness of sense, its way of continuing to come and to be on its way, without confounding it with an indeterminacy that lacks all consistency? What name could open up an access for the anonymity of being-in-common?" (Nancy 1997, 90). This is where poetic speech and the sounds of the crowd play an important role. Regarded as "noise," such speech does not solidify into a, or the, "truth"; there is no ownership of a "message," nor is there dominion over the space of its own resounding. Instead there is a habitation and exchange that occurs within a shared environment. Note that "rumor," "murmur," and "hubbub"—all terms for crowd-sound, describe a collection of voices *as* aural communication that lacks structure, where the echo of the repeated syllable both reinforces the sound-minus-language of onomatopoeic, structureless speech and reaffirms through the repetition of the second syllable its questionable existence. Noisy and poetic speech, like the wind, like air itself, is open to mutation and resists fixation—to a speaker, an object, an "a." In avoiding individuation, however, it also reminds speakers that their speech is forever under appraisal, that they are not so far from the oppressive controls associated with childhood, insanity, or senility.

The paradox of noisy speech, its lack of coherence into a "message," is, however, exactly its power. It is, as Nancy describes, a vehicle for the transmission of sense or even the possibility, the ground or condition, of such transmission. In this respect, it functions as "art" in that it "opens sense" in a way that is very similar to the sharing of a secret within a room, signified perhaps by an embarrassed silence:

> There is "art" every time a sense more "originary" than any assignment of a "Self" or "Other" comes to touch us. This can take place in gestures, postures, the "art of conversation," social convention and ceremony, rejoicing, and mourning. ... This does not mean that "art" is everywhere and without distinction: "art" is merely that which takes as its theme and place the opening of sense as such along sensuous surfaces, a "presentation of presentation," the motion and emotion of a coming. (Nancy 1997, 135)[7]

Clearly, such moments are not "art," which is to say that Nancy's definition of art is unorthodox, outside of the doxa that sets art apart from the mundane world from which it arises. We can imagine what it would be to re-insert this shared experience of sound as the passing of wordless and unspoken knowledge—secrets in the traditional sense—of that which is obscure, enigmatic, and incomprehensible. A good example is the moment after the orchestra has finished, when the symphony has come to an end, when the violin bows are still raised in the air and only just beginning to fall down into the laps of the players—the moment when the audience realizes that this is the final movement. In that brief pause there is a resonance chamber of silence during which, if the orchestra has played particularly well, if the virtuoso has stunned the audience so vigorously that they are paralyzed, the last reverberations of the final chord will flood the audience in a quiet diminishing echo, within which speech is excluded. For a moment, a wall is erected against speech. After such a pause we might hear someone whisper, "Sublime!"

This is the transformation of noise into an expression that registers consensus, a shared understanding, a sympathy, empathy, and one could say resonance, and the short-circuiting of language and discourse to produce direct action, sometimes a standing ovation. This is a power that audiences take upon themselves in their response to the performer. But in order for this moment to occur, a moment of speechless appreciation, transmitted, or rather reverberating, resonating, flowing through the crowd from front seats to back, from inner circle to the pits, there must be a hall, a mass of bodies, an atmosphere, and the possibility for a moment of time. Without this architecture, there can be no moment of silence after music and before speech.

Serres writes that the knowledge that comes through this kind of silence should be treasured, describing it as the transformation of one acoustic system into another. What is it a knowledge of? In this context, it's more like an experience of collectivity—an unspoken acknowledgment that there is a harmony between the orchestra, the sound, and the audience. It is the same experience that allows the crowd to laugh simultaneously at something unspoken, for people in a conversation to nod in agreement although nothing has been said, for a silence to descend upon a congenial gathering as a recognition that something unpleasant has occurred, for people to finish each other's sentences. Some kind of knowledge or understanding passes between people in the absence of language and manifests often in a sigh, a nod, a change of posture, a departure, a raised fist, or a mobilization. It could be described as a sense of attunement—a *stimmung*. But this would not be an attunement of the soul or conscience or the self, but rather a mood or atmosphere, an "inclination" or "disposition" that is fundamentally communal. For Nancy, it is a disposition "in which we acknowledge neither the origin of communication nor some transcendent source that legitimates all communication … [rather] it defines our very existence as existence *in the world* rather than returning us to an interiorized essence, a self-contained subjectivity" (Armstrong 2009, 118). In other words, the sympathetic resonance that occurs in (sonic) attunement does not reference a Heideggerian return to the innermost being or any form of interiority.[8] Rather, it extracts from the already-interiorized connotations of "mood," the active, moving, and fleeting sense of e-motion: a motion that cannot necessarily be caught, named, owned, or adopted, but circulates, and is passed around in a form of commonality, or, as Nancy would say, "being-in-common."

Poetic speech shares this kind of consensual practice and inflects everyday speech with the recognition that sometimes the only way to express something is through "noisy" speech—speech that is more like vocalization or babbling than language. This may be the form of "music" within language that Serres is referring to when he writes: "Meaning presupposes music, and could not emerge without it. … It inhabits the sensible, it carries all possible senses. It vibrates in the secret recesses of our conversations, continually underpins our dialogues, our exchanges presuppose it, it knows in advance harmonies and discords … and [it has] paved the way for our collective existence" (Serres 2008, 123). We forget that language happens in sound and that meaning is the tail end of a far more profound conjunction. Writing, even philosophizing, is motivated by the rhythms and intonations woven into meaning. In so forgetting, we believe that we understand a voice without resonance—or, more frankly, a voice without sound.

A Voice That Is Not a Voice

music is how
we first remember
each thought carries the ruin of a sound
the mouth
breaking the surface

—Amanda Stewart, "phoneme" (2010)

The murmurings of language, the tones of discourse, the uncertain space between phonemes, the guttural power of the voice as "ethos incarnate," and the "voice-that-is-not-a-voice"—these areas, or topoi, have and continue to be investigated by Amanda Stewart, an internationally recognized poet working across multiple artistic fields.[9] Over the past two decades, Stewart's work has evolved from what would be regarded as voice poetry to sound poetry, and, most recently, noise works. Stewart describes her earlier work in the 1970s and '80s as essentially discourse analysis, spending years as an audio artist, author, and filmmaker deconstructing political discourse and especially the language of the nuclear industry:

> I wanted to understand the oral aspects, the rhythmic structures, the source of vocabulary used, the tones of voice, the whole layers of "being-ness" that emanated from that discourse and was then transferred and re-spoken in different ways by the culture that (from the 1940s onwards) was starting to assimilate it. In the '80s I became increasingly interested in the links between distinct properties of different written, oral, and electronic forms of language and the "forms of inscription" or "modes of memory" that they engender.[10]

Stewart's focus then shifted to the question of language and the production of subjectivity through the voice, which, "unique amongst instruments, has the power to synthesize musical, semantic, psychoanalytic and emotional structures, within one entity." Using multitrack, she would layer text, often having parallel texts, including a series of pieces that explored subject-object relations in language. "There was the object text and the subject text, and then two layers of stereo improvisations that commented on those two subject/object texts." Stewart would break sounds and language into phonemes, and, through stereo miking, project the sounds into the back of the performance space to create a distance between herself, as producer of the sounds, and the voice that she heard. This distancing enabled her to "scramble" the existing codes that would normally attach a phoneme to meaning. The degree of difficulty involved in shifting between two microphones (panned left and right) and multiple texts sparked an interest in the effect of "disassociating the voice from the speaker using stereo ... its stereo projection away from you, into a sculptural materiality, that can actually return the self." This "self" is, however, not the self-same self of the classical speaking subject.

> Well, "self" isn't exactly the right term [since] you experience a split.[11] It's a disorientation but also a reorientation [and at the same time] you're opened up to all these nuances in language that you normally wouldn't experience, so it becomes a revelatory process. I found that I began to break down language and started using a lot more extended vocal techniques and experimenting with what happens when you break a word down to its sonic elements and then start extending and stretching them ... and this eventually led me to start experimenting with the idea of making a voice "which is no voice," a voice which, if you couldn't see that there was a body there, you would think it was an electronic sound, produced by some kind of digital processing.

Despite the difficulty of creating "a voice which is not a voice," Stewart has persisted for some years, motivated by three overlapping desires: first, to rid the voice of signification, to counter the voice's power as "ethos incarnate" (Stewart 2010, 174). According to Stewart, the power relations involved in discourse create a flow between the "pneumonic and the mnemonic," the one regulating the other. These mnemonic systems are multilayered, complex, interconnected, and physically manifest in ways that we're all familiar with, such as when our body stiffens, we sit upright and shout "Don't say that!" or bow our head, hunch over, and say softly, "I'm really sorry."[12] Her more recent noise works amplify this somatic effect because in order to make the sound, Stewart doesn't use her body as a resonating chamber but rather, "I'm pushing the breath up through my mouth and then manipulating it up around my mouth and my teeth in order to get these high-pitched or deep sounds." These are the sounds that Stewart produces in order to achieve her second objective—to fashion sound/noise that resembles "pulling the plug in and out of a socket so it makes short, static-like sounds" that would blend in with the digital sonic envelope emerging from the laptop improvisation ensembles she performs with. The effect is to completely mask her voice such that the audience, her collaborators, and sometimes even she, herself, don't know who or what has produced the sound.

With the increasing splintering and dissection of sounds ("micro-sounds") in computer music in the late '90s or early 2000s, Stewart began to feel that her vocal presence within certain improvised music was too obtrusive, that her vocal techniques were unable to match the short, micro-sounds of, for instance, the synthesizer or the computer. This led her to develop a series of solo noise pieces composed of extremely short and vaguely electro-sonic sounds that curtail or "hold" the breath, releasing it very quickly and surreptitiously (in order not to "pop" the microphone, which would indicate the presence of breath, of a body, and ultimately a vocalist). Indeed, disguising the presence of the body and a voice meant that Stewart had to remain motionless, her body tensed and hypervigilant with regard to her breathing.[13] Otherwise, as she explains, "My voice would be recognized as voice" and intrude upon the sonic envelope already established by the other performers and the computer-generated sounds they produced.

This has meant developing techniques that mimic, to some extent, the sounds found in contemporary noise works, not just in terms of sonic content—for the frequency range and continuous sounding of electronic instruments are impossible to reproduce—but also in terms of the organization of the material among the performers, which is in part influenced by the instruments they use: "For example, with wind instrumentalists—unless they're going to circular breathe—their gestures are going to be determined by breath. Having a computer frees you from that." In a peculiar irony, Stewart has persisted in using analog technologies rather than real-time digital signal processing in order to produce a voice that, in certain circumstances, has a timbral awareness of its analog nature *within* the field of the digital.[14] Being able to blend in with the sonic envelope of laptop improvisation also achieves Stewart's third objective, which, influenced by the ethos of improvisation as a democratic process, would remove her from the traditional role of the vocalist occupying front and center stage. Dismantling the hierarchical structure of music performance corresponds also with the kind of "music" being produced. Stewart mentions the history of electronic sound arts as one that has frequently been motivated by the desire to eradicate both the "star" performer in a concert or performance and to liberate the performers and the audience from the discourses and codes that shape our listening. Laptop performance establishes the performer as one who simply directs sometimes preexisting, sometimes improvised and "live" sonic transformations and sonic "objects." The computer becomes an "instrument" but not one that would evoke the virtuosic technicity of, say, the violinist or the pianist. There are a whole range of sounds, discourses, identities, and listening expectations that are unsettled in the situation, but one that emerges frequently is the assumption that improvisatory noiseworks—especially those produced by computers—are, by nature, able to eradicate the individual's subjectivity, to create an environment of equality among performers and audience, and among performance-based sound (what would normally be called "music") and the sonic world. Stewart had already been experimenting with disassociating her voice using multitrack and stereo miking and wanted to try something completely different from the "subject/object pieces":

> Instead of doing something that's picking up on all the internalized grammars and pitches and breath structures, and all the complexities and residues of what comprises language, instead of playing with that level of disorientation of having multiple oral and written texts, and the different techniques that I used to confound me as a subject. I wanted to see if it was possible to strip signification from the voice, [to bypass] the automatic reaction of interpolation, where as soon as you hear a voice your subjectivity is affirmed and restored.

However, in mimicking the freedom from the rhythms of breath that computer music allows, Stewart has to adopt a particular stance—one she

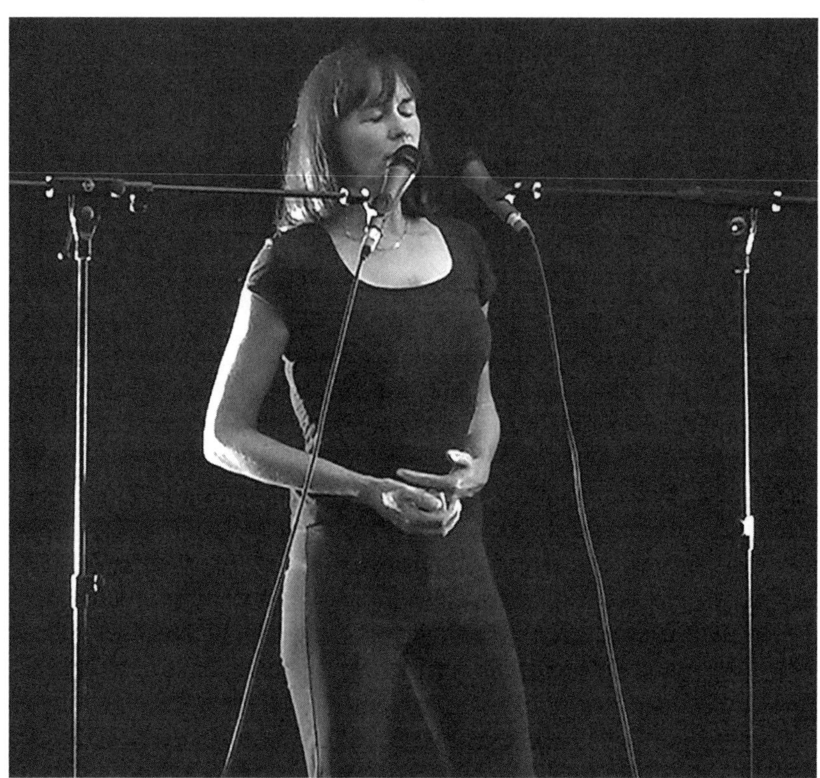

Figure 5.1
"As If," performed at Artspace, Sydney, 2007. Courtesy of Wade Marynowsky.

describes as "grotesque." "When I am doing this thing you could see my right hand like a claw and then two of the fingers moving slightly, almost like twitching. I just looked like Frankenstein's monster!"[15] The posture and stillness required for Stewart to perform her noise pieces coincides with the disciplined composure required of the audience to fully appreciate, through intense listening, much of the sound art and experimental music performed in the unused or retrofitted warehouses and industrial spaces that have, in Sydney at least, become the principle venues for such work.[16] Without wishing to generalize, she notes a tendency in the '90s and early 2000s for performers to focus on "the idea of extended tone, just using a mixer with feedback within the mix without any inputs or outputs." In order to hear the very gradual modulation of tone, its often barely perceptible changes, not only does the listener need to quiet his or her body, but the acoustics of the entire space need to be taken into account in ways that are simultaneously inclusive and adaptive.

Stewart's approach brings into relief the difference between local, "unorganized," and sonically diverse sound events and performances, as well as the homogeneity of what I am calling "institutionalized" sound. A very good example of such institutionalization is the sonic envelope that is often created in gallery-based sound installations. For instance, a recent exhibition at the Museum of Contemporary Art Tokyo, "Search for a New Synesthesia" (October 27, 2012–February 3, 2013), demonstrated the need to organize the sound artworks and installations according to the acoustics of gallery space, which was highly reverberative.[17] One piece in particular was exemplary in this regard. Otomo Yoshihide and Yasutomo Aoyama's *without records* (2008) occupied a huge, three-story, granite-walled atrium, with various instruments filling the space from wall to ceiling. Described by Yoshida as a "forest" of old portable record players, the turntables were "prepared" with various bits of metal, plastic, and rubber. The 119 turntables were activated by a computer-controlled sequence that randomly triggered their operation. Listening to the spatialized and sonically diverse sounds, one could almost imagine a mechanical forest, with the generation of sine waves providing an aural coherence, smoothing out the jagged edges of the mechanical sounds. A sonic ecology, if you like, that existed within the gallery and yet did not seem to "bleed" or territorialize the other spaces—housing, for instance, the delicate sound of bowls bumping against each other in a swirly pool of aqua water in Céleste Boursier-Mougenot's *Variation*. The reason, I discovered, was the sonic envelope of the exhibition itself, which allowed the visitor to move from one piece to the next without a radical change in the acoustic timbre or shape, without wild fluctuations in frequency ranges or volume. Ikeda's *Datamatics* was perhaps the exception, exhibited in room as a series of separate screens on the far wall, projecting the different data streams that comprise the piece, and fronted by separate podiums upon which headphones were placed for individualized hearing. Note that this presentation contrasted sharply with the performance of *Datamatics* at MOFO in Tasmania (2012) discussed in chapter 3, where the sound and image engulfed and palpated within all spaces available, including the listener's intestines.

Allowing each piece its "space" was in this case, and many others, also manufacturing a particular overriding sound. The same could be said for the sonic environment in which Stewart performs—and more broadly, the sonic envelope of digital music. According to Stewart, "Digital technology

... profoundly affects and alters our pneumonic system and mnemonic systems even when we don't use that technology" (Stewart 2010, 174). In conversation, she mentions the short, fast-paced, clipped voices of media personalities that exemplify the technics of digital media:

> There's a certain speed of speech ... it all has a squashed in boxy linearity to it—there's very little depth, very little resonance. ... We're surrounded by compressed sounds with everything upfront—in your face, so you don't have the ability to lose something, to not hear something, everything is "there" for you, all the information is compacted in the same band—it's all rapid fire and always in a linear information flow.

Practically, this means that Stewart has to adapt her vocal performances to the exigencies of the microphone and recording setup so that her voice also has a "bright, upfront sound." The full significance of a compressed and breathless voice is beyond the scope of the present discussion, and I would refer the reader to Jonathan Sterne's excellent account of audiophonic compression in *MP3: The Meaning of a Format* (Sterne 2012).

But it is important here to mention that the perceptual encoding he discusses, having habituated the listener to this sonic quality has also eviscerated the space, or time, between words—the "unsaid" that is equally important in Stewart's work:

> What is not uttered is multi-present with what is uttered. In the complexity of the human body, sound, meaning, and emotion are actually synthesized and latent, and in fact the breath is one of the openings for that latency to come out, and that's what gets sealed off more and more with compressed/media sound. [The media voice is an] unrelenting, compressed, all-knowing voice, a voice that has no self-reflection, no ambiguity, it's sealing off the present, it's patching and fixing as it goes. There are no pauses, no time for the listener to digest.

Whether in a gallery, from a television, or online, the sonic envelope *envelopes*, shaping the culture it also (re)sounds. We need to ask, what takes the place, or fills the vacuum, when the breath is shut out? What occurs, or doesn't occur, when there is no pause in speech? What does the constancy and continuity of voice do to a conversation? What happens when the speaker "never takes a breath" or is "longwinded." Returning for a moment to the role of acclamation discussed in chapter 2, if the eternal repetition of the Sanctus continuously reaffirms the existence of God, it also obliterates the possibility of a silence that could, for a moment, allow the heavens to fall and the angelic hierarchy to collapse. The continuous voice saturating the twenty-four-hour media newscast forecloses the possibility of a dialogue, reinforcing the power relation inherent in a voice that has no resonance, that allows no space for reflection, and that substitutes actual acoustic reflection that would occur in face-to-face conversation with a monotonous and didactic stream. Whereas in one-to-one conversations the person standing next to you or opposite also serves as a literal sounding board, returning your voice, and thought, allowing ideas to form, debate to arise, ideas to be contested, these possibilities are disabled in the absence of the time and the space of breathing.

Poetic Speech, Power, and the Anechoic

Stewart is sometimes referred to as a sound poet (although is more correctly described as a poet working in the area), but she eschews the notion, common in the sound world, that "nonverbal vocal sounds are nonsemantic" because for her, "as soon as your voice is 'a voice,' it's semantic." But to create a voice-that-is-no-voice is—taken to the extreme—an exercise in self-annihilation. In this respect, Stewart's work exemplifies the paradox of sound's ontology, its essential nonbeing as a *unit* of being, and its role as the sonic equivalent of what Žižek refers to as the "symptomal real": both that around which a particular situation is structured (i.e., its foundational term) and "the internal stumbling block of an account of which the symbolic system can never 'become itself,' achieve its self-identity" (Hallward 2003, 90). Stumbling, like stuttering, implies a temporary interruption in a process that would otherwise move toward completion. The refrain of the Sanctus, the onomatopoeic echo, or the silence before applause functions poetically only within the context of a final end point or termination, a teleology that is theistic or metaphysical. The failure to reach an end point is something like the voice caught in the throat—it simultaneously suggests an inability to incorporate and the stubborn resistance of the corporeal; the annihilation of a self that no longer fits a technological regime, that is also an affirmation of a voice in whose throttled sound lies a degree of freedom *to* speak; an articulation, a stuttering that, as Nancy writes, "betrays the form of the problem: 'we,' how are we to say 'we?' or rather, who is it that says 'we,' and what are we told about ourselves in the technological proliferation of the social spectacle and the social as spectacular, as well as the proliferation of self-mediatized globalization and globalized mediation?" (Nancy 2000, 70).[18]

In naming a beginning and therefore pointing toward an end, the big bang reveals, in its onomatopoeic simplicity, a startling feature of contemporary knowledge production—that noise in language (poetic speech) substitutes for what either cannot be said or heard, or what we refuse to say or hear. For not reaching an end, having no reflection is somehow a reversal of the constitution of power—divine or diabolical, it suggests an *aporia* without structure—an anarchy. Finitude is part of the constitution of power; however, finitude also has a literal sense. Thinking of sound as a spatial phenomenon, this necessary finitude presents itself in the form of a physical enclosure. Thus we find that sanctified enclosures such as churches or concert halls affect the transformation of "noise" in language: from being heard as inferior to language (the repetitions in onomatopoeia), to being beyond language (the repetitions in the Sanctus), from a kind of animal cry (acclamation), to an imitation of celestial worship or discourse with divinity. The interpretation of this speech depends entirely on its context, specifically its architecture. What kind of enclosure gives this pseudo-speech its status? Even if only a loudspeaker, the voice has to be contained within the physical symbols of power. But this is not to say that poetic speech is "excluded." On the contrary, within prescribed domains (poetry, media) poetic speech functions like an echo—to prolong power, intensify it, multiply it, and cover up its finitude, filling the void that is its core. Media echoes, both sonically (filling acoustic space) and atmospherically (filling bandwidth). Its saturation, its presence, and its power are not just sonic but represent a global system of substitution: media presents a replacement for actual presence, representative voices—be they political or media—a replacement for actual voices, the IPO (initial public offering) a replacement for the "public," and more recently, the market (or to be more precise, bond yields) a replacement for democracies. The paralytic response to some of the world's most pressing problems, or rather the complete silence regarding such problems, is masked by the quiet and constant roar of media, the dense networks of communications culture creating a forest of signs, a canopy of beeps, texts, chats, tweets, calls, grabs, bites, blogs, and the array of networked communications that are on hand to shield us all from the warming sun. We can think of these institutions: government, finance, and media as vehicles for modulating the dissatisfaction and distrust felt by a large proportion of the populace into something that resembles that dissatisfaction but is eviscerated of all demands, all outcry, all fury. This muted response returns to the ears of the populace as echoes, faint simulations of their will, from which little sense can be discerned. Indeed, the ability to sense, to make sense, and to inhabit the sensible has, as Serres writes, been suffocated in the complex movements of noise and sound through the atmosphere, the body, the media, and back again, in a cycle of increasing material consequence, yet decreasing human awareness.

In this context, the transposition that occurs in *Listening Post* part 1 mentioned in previous chapters—from the Orwellian connotations of intercepting and tracking Internet chatter into the far more benign reading of a system that simply "speaks the world's thoughts"— is instructive. For this transposition to occur, the hard associations of surveillance must be ameliorated, the noise of social chat, produced in private rooms and subject to sticky taboos, must be musicalized, the harshness of chatter resolved into chant. This process contrasts with, for instance, Ikeda's *Datamatics*, where the "wetware" of genomes and the cosmological figures of planetary constellations (elements that are measured within coordinates, targeted through crosshairs, and identified by beeps) involve a process of ordering and hardening. Following Serres, we could describe *Listening Post* as a set of boxes through which energies pass and transformations occur. This schematic begins at the cellular level and moves in continual cycles through to the social. (As we shall see in a later chapter, the new financial instruments—"boxes," if you like—that have monetized the social represent an additional cycle.) According to Serres, the first of such cycles begins with the cells, vibrating within the body that are then calmed by the external world. The buzzing of the cells moves into self-knowledge, to the cry of pain for instance, followed by the words "I am in pain." This process is regulated by the force of sensation, soothed by "the hard sonorities of the earth": the gushing of water, the rustling of leaves or the screaming of gale-force winds (Serres 2008, 118). Small sounds, high-frequency energies, are quieted by the low, slow frequencies of the movement of the elements. These are the sounds that, according to Serres, belong to "the given," the material world, the world that we sense. They pass through the body and are filtered by the skin, and in this way, noise and vibration become meaning, hard sounds become soft, and the world becomes bearable. Serres stresses that if we didn't have these filtering devices, the world would overtake our tissues, in short, that if we didn't have an eardrum, we wouldn't have a brain. These membranes, the skin included, protect us from the elements. But protection can go too far: clothing, houses, air-conditioning, insulation—each one blankets experience, creating layers that fade easily into the background hum of existence, and with their withdrawal, so too the recognition that the small amount of light that enters through windows or the narrow range of frequencies that we allow to vibrate within us actually speak for the world.

This is the first cycle that Serres elaborates. The second cycle is defined by, and defines, the social. The social is the epicenter of transmission-reception, of the filtering processes that transform hard sounds, coming from the sensible, into the soft sounds of language, symbols, and mythologies. Serres describes the social as "an immense, transmitter-receiver, social box."

Through the social, the given is continually refined, abstracted, and softened, but at the same time, muted and deadened. The noise of the collective as a *racket*—drowning out the body, silencing the world, with its endless

> shouts, car-horns, whistles, engines, cries, brawls, stereotypes, quarrels, conferences, assemblies, elections, debates, dialectics, acclamations, wars, bombardments, there is nothing new under the sun, there is no news that is not news of yet another racket. Noise is what defines the social. Each is as powerful as the other, each multiplies as quickly as the other, it is as difficult to integrate into one as into the other, the transition from chaotic rumbling to information—no matter if it is meaningless provided it appears organized. (Serres 2008, 107–108)

This "thunderous flux" forms the social contract, which is continually renegotiated through cycles from hard to soft sound, from a sense of belonging to feeling isolated and alone, from one medium, one space, to another: Facebook establishes one experience of collectivity by displacing another—conversation at the dinner table withdraws as the quiet tapping of screens, gasps, mutters, exclamations, and expletives replace the louder sounds of voices talking. While the social contract unifies the group transmissions, at the heart of unification is disassembly.

In the second cycle, these filters become systemic, are refined, narrowed, analyzed, and filtered again until we see and hear only our filters, believe only our rhetoric. The vestibules of the skin and the ear are not touched by the movement of sensation, the movement that produces sound in the first place. They receive instead a different form of energy, transmitted and processed via media that have their own relation to atmosphere, through which pass voices that are not "voice," motivating a hearing that is not hearing. Adopting the full force of Serres's and Nancy's notion of listening, it could be argued that what is heard through an audio speaker is such a reduced and diminished form of sound that to call it "sound" is a misnomer. The diminution of media sound is only partly acoustic. What is more pressing is the tacit agreement that conversation is actually occurring despite a decoupling of voices, bodies, and physical space. Conversation, dialogue, and debate become a venue for the circulation of slogans, a racket that supports the racketeering we will discuss later. The effect on sense is multifaceted: constantly hearing media voices that fail to resonate, in all senses of the word, reduces the dimensions of meaning, replacing it with a kind of media-induced dementia. The solipsism of endless iterations of transmissions that circle the globe with ever-increasing speed can be heard as the tinnitus-like ring, the very high-frequency pitch, barely discernible, of electronic communication. A lack of redundancy perhaps, a lack of openness, the aural sense of myopia, produces meaning as a form of movement, the reiteration of the same fragments of speech circulating so fast and with such vapidity that it is literally a form of aural "spin." Most of what is said is gibberish—the back and forth of arguments and discussions creating a racket that is not just noise but also a kind of currency, an unacknowledged trade maintains the status quo in its circulation.

The third cycle transforms the inaudible buzzing of the cells into the racket of culture and finally into the soft, quiet, and calming presence of a street sign, a billboard, map, speech, or declaration. Serres calls this a "white box," within which the sounds of the body and the sounds of the social are transformed yet again, manifesting as signs and symbols, transforming the circulations—from the inaudible buzzing of the cells, the thunderous noises of culture—into another mode of meaning, but at the same time taking the edges off, introducing more filters and more regulations. From "energy to information," from things to signs, from the harsh sounds of physical labor with the muscles working in unison—banging out metal in the forge for instance, shaping it into the silent relations of tonal harmonies that orbit the music of the spheres. Within the high-frequency buzz of the networked collective, the cycles from noise to sign quicken while the dimensionality of meaning or tone drops out as a kind of material residue. The belief that, as Serres says, "the world is crisscrossed by nothing more than signals" creates an accelerator, where the movement from hard to soft, from noise to sign, becomes a surrogate for sensation: the skin vibrates not to the hard sonorities of the given, but to the spectatorial lightness of spin. Increasingly, the social racket, insulated by language, ensconced in its various enclosures, and believing its own rhetoric, obliterates the multiple dimensions of meaning. What is lost in the process is, according to Serres, "quite precisely—our common sense." And by this he is referring to both the sensible (as all that can be sensed) and the idea of a common wisdom formed from the union between the sensible and the "generalized eardrum of our skin."

Sensation is heralded by noise—the noise of movement, not of things. Sensation flickers on the skin and gathers up the body, pulls it out of its tempestuous inner monologue, and presents it with the world. This may be the only dialogue that we can have with the given: a dialogue of movement and ephemeral sensation. But this is not to say that there is no communication: the gentle gusts of wind create a union between the body's tissues and atmosphere. This union has a voice that resounds in our voice as a residual undercurrent, moving beneath language, as a kind of music, and beneath music, as noise, giving the voice meaning and depth.

For Serres, knowledge is moved and circulated by the tonality of speech, which is itself tethered to flows of sound. Without this essential tone, he writes, "Eloquence ... collapses into gibberish and boredom" (Serres 2008, 22). We find the anechoic house becoming a reverberation chamber that contains and controls its inhabitants, with endless reiterations of the mythologies forming between us. Reverberation in an enclosed space obliterates foreground and background—all sound, all thought, blends into noise. It is impossible for a voice to have resonance in this space, and without resonance, the voice has no meaning—no rhythm, no tone, no sense.[19]

In a dynamic that Serres admits is almost impossible to unwind or reverse, the echo chamber of media not only circles but also mutes the world. But this muted world now thrusts itself "brutally and without warning into our schemes and maneuvers" (Serres 2008, 3) just at the time when our awareness has become global. "Empiricism is not enough to wake us from this new sleep, we need an eruption, a large-scale seismic event, a major cyclone, a new Hiroshima. But no: the ocean rises up on our screens, voluptuously" (Serres 2008, 114). It may be that "the ocean rising on our screens" Serres is referring to belongs to an era that we either have passed or are approaching the passing of. Of course, the sounds and sights of the earth are accessible via various screens—they can be experienced precariously and often these are the only experiences possible. But Google, and all the other technologies of mass surveillance propagating the Internet right now, creates its own enclosures. The blue planet is a screensaver that may seem innocent in its ubiquity and universalism but is also the grotesque output, or merchandise, of a filtering process through which "the sound of cannon-fire gradually transforms itself into a whispered confidence" (Serres 2008, 115).

This process is, as I will later discuss, one of the main forces shaping our current economic, physical, and psychic climate. But rather than cannon fire, the weapons of choice are sophisticated financial instruments, engineered to take advantage of "market noise" in much the same way as surveillance has been engineered to take advantage of Internet "chatter"—that is, to operate through complicated algorithmic processes in synchronicity with equally complicated legislative, political, and media-rhetorical processes. In *The Ecological Thought,* Timothy Morton describes these processes as a "furious yet ultimately static whirl," remarking that "we find it easier to imagine the collapse of the Antarctic ice shelves than the collapse of the banking system" (Morton 2012, 19). A whirl creates the impression of movement and therefore sensation—as if dialogue is going somewhere, as if debate is escalating to a point where something or some action may occur. The escalation, the back-and-forth debate substitutes for actual physical sensation, and in its substitution creates a habitat, a surrogate environment, cultivated from the nonspace of the market on the one hand and governmental institutions on the other.

Sonic Monotony

It is important to remember that the anechoic is not just an environment for eliminating reverberation, a form of sound insulation, noise reduction, or even sensory confinement. It is also the product of excessive sound—of an acoustic environment saturated with competing layers of voice, sonic signals, ambient sound, and noise. In the context of mass media, a common and persistent critique is the presence of an underlying drone that sonically privileges the very loud voices being heard (those broadcast, those financed). However, what is new to the millennium is the dynamic combination that the anechoic represents: sound insulation, media obliteration, and the deracination of sense through the multiplication of the fracture, or separation, between ethics and practice, that dulls our ability to fully appreciate the crisis at hand. These are not three, separate effects of the anechoic; they interact as a dynamic system. As Serres argues, enclosure from the world contributes to enclosure of the world; stifling the senses also diminishes the ability to sense, and that is also the ability to make sense of the sensible. The horizon of sound in need of consideration, however, includes sound and all that sound implies: all that it carries, both forward and backward (knowledge or an indication of its origin). For instance, "a huge sound, it made me jump," imparts the distance from the listener, the atmosphere that the sound is moving through. "I could barely hear it, the rain was so heavy"—the clarity of the sound as an indicator of the thickness or thinness of the atmosphere, the amount of humidity, and so on. Also, knowledge of what will happen next: the very first creak of the tree as it begins to fall we know will be followed by a massive thump, which will be followed by a shaking reverberation. These sounds and this kind of listening contain within them a form of knowledge about the environment and a mode of sensing the environment determines the process of knowing and the skill of listening itself. Constant attention to only media sonority is a deskilling process. The monotony that it produces is a form of sonic air-conditioning.

Indeed, air-conditioning provides a model for this dynamic and a perfect illustration of the "anechoica" under discussion, producing what Peter Healy describes as a form "thermal monotony" that has become standardized across the developed world.[20] Healy traces the development of air-conditioning as a standard, similar to plumbing and electricity, that became integrated in the design of housing, regardless of the climate, and had effects on the structuring of space within office developments and other commercial spaces. Since windows were no longer needed for ventilation, natural light could be replaced by fluorescent lighting, contributing another element to what has become essentially a new atmospheric and experiential dynamic, creating its own microclimates—both indoor and out. The incessant hum and pulse of fluorescent lighting established a sonic norm, institutionalizing a level of mechanical noise within domestic and commercial environments. Schwartz notes that

> architects who used double-paned windows, sound-absorbent panels, and acoustic ceilings to cope with the noise of typewriters and ringing phones in the open-plan offices that arose during the 1950s would by the 1970s use heating and ventilation ducts to add in white noise as "sound curtains" veiling desktop clicking and as "acoustic perfume" to give office workers a shared backdrop of productive, communal activity. Was the blow and hum of air conditioning units perfectly suited to the (white noise) job of masking unwanted sounds in apartments with low ceilings, weak walls, and thin windows, or was its "metallic threnody" an omen, as Time heard it, hanging "in the air above close-nestled, rich communities like the thrum of some giant insect infestation"? (Schwartz 2011, 838)

If, in the '70s, the white noise of air-conditioners canceled out the more brutish click and clang of typewriters, in the new millennium the sound of the air-conditioner calls for even more soundproofing. Both sonic and thermal volume are now integrated: they reciprocate as candidates for insulation against, on the one hand, and production of, on the other. In all respects this is a ludicrous exercise, with one canceling out the other in the quest for thermal and sonic homogenization. As the air-conditioner cools the indoors, it increases the volume of heat and sound in the immediate environment outside, creating a demand for insulation in the neighboring apartment or office block. Yet one simple division ensures that the cycle of production and negation, based on what is essentially a filtering of the elements (light, heat, and sound) that make up sensory experience (vision, touch, and hearing) and that is the division between inside and out, the separation of space between entities (be they individuals, land masses, or corporations), and the proprietary demarcation of, and control over, space.

The conditioning of the senses and the atmosphere is also a conditioning of "common sense." The standardized, insulated environment and its correlate, the acoustic and thermal monotone, affect the capacity to sense and also the capacity to think. In this way, controlled spaces and conditioned senses pass from the sensual to the aesthetic, the social, and the political, from hard to soft, from discourse to sign, and now to algorithmic formulae. "Speculation," that eminently philosophical term, has little to do with thought, and everything to do with finance. The reduction of resonance, reflection, and the returning, reverberating voice to a mere repetition creates a massive, reverberating echo, and in this context, the myth of Echo is prescient:

A strange-voiced nymph observed him, who must speak
If any other speak and cannot speak
Unless another speak, resounding Echo.
Echo was still a body, not a voice,
But talkative as now, and with the same
Power of speaking, only to repeat,
As best she could, the last of many words.
(Ovid 1986, 61)

In this myth, Echo has no resonance with her object, for her voice is rejected. "Her body shrivels," all that is left is voice and bones, and eventually her bones are turned to stone. The doubling of sound is a reminder of rejection, the voice is reduced to "just a sound"—"alive, but just a sound." A voice without body is "just a sound" in the same way, a reflection from the hills or from any surfaces is not necessarily an adulation, as in the repetition of sacred verse. The place of sound, where "sound resounds," isn't just an empty place, a space of nonbeing between the reflective surfaces that produce a reverberation. Rather, it is the nonsubjective, nonmaterial ground *and* instrument of individual and social formation.[21] If homophony unites the two ecos—economy and ecology—it should also ground these in the reverberations of human (sonic) emissions. It is here that the multiple meanings of "sense" emerge, for sound is as much a phenomenon that is heard and felt through the senses as the main vehicle of the production of meaning. This occurs through discourse and primarily through the voice. So we can think of "eco-echo" as a dynamic system that encompasses natural phenomena and cultural production, within what has obviously become a finite setting, but for centuries has been regarded as infinite. The bounds of finitude bring the two ecos together: growth cannot go on forever, but the dynamics of the echo—especially in the media sphere—ensure that this simple fact will be modulated, distorted, and reverberated beyond sense and ultimately beyond hearing. As Echo was cursed with the monotony of repetition, voicing only the last words of men who spoke, the echo-sphere circulates fragments of authorial voices as endings, terminations, leaving nothing more to be said.

Echo is the modern-day form of acclamation as public opinion, "multiplied and disseminated by the media beyond all imagination" (Agamben 2011, 276). Echo is, however, not so much about the return, reverberation, and reflection of public opinion, or even its management, but, according to Agamben, is concerned with the transformation of government itself into spectacle, one that performs a ritual of governance for the glory of finance.

> The function of acclamations and Glory, in the modern form of public opinion and consensus is still at the center of the political apparatuses of contemporary democracies. If the media are so important in modern democracies, this is the case not only because they enable the control and government of public opinion, but also and above all because they manage and dispense Glory, the acclamative and doxological aspect of power that seemed to have disappeared in modernity. The society of the spectacle—if we can call the contemporary democracies by this name—is, from this point of view, a society in which power in its "glorious" aspect becomes indiscernible from economy and government. (ibid., xii)

The correlation between religion and politics is in many ways quite obvious: a prayer is said before Parliament or Congress sits, the parade into Parliament performs many of the rituals associated with the mass, and national anthems often mimic the structure of hymns. However, this correlation, while important in itself, is less profound than the point Agamben is making: that this correlation is possible, that religion and politics can exchange robes, only because

> underneath the garments there is nobody and no substance. Theology and politics are, in this sense, what results from the exchange and from the movement of something like an absolute garment that, as such, has decisive judicial-political implications. ... This garment of glory is a signature that marks bodies and substances politically and the appellate theologically, and orients and displaces them according to an economy that we are only now beginning to glimpse. (Agamben 2009, 194)

In the same way that governance has been installed historically through the amalgamation of theological power and political power and through the separation in the former of absolute power divided within the Trinity and then transmitted through the deity (through the process of division and delegation discussed in chapter 1), governance is currently installed through arcane financial processors that distribute debt (as a division of the negative) to the point where the coherence of the financial systems has been shaken to the core. As I will discuss in the following chapter, the "spectacle" commonly associated with media can be seen as the visible cloak of (financial) speculation which, in its glittering complexity, has produced a debt economy that is truly unfathomable, and has become the modern day "mystery of the economy."

Like Echo, the function of this financial process is to reiterate, reverberate, and endlessly "spin" off tranches of derivatives and debt obligations until the algorithmic bubble bursts—or not. Echo intersects with economy at a level of technological/mathematical infrastructure, which removes the habitable sense of *oikonomia* as home, leaving only the administrative function, now operated by transfers of signals that are imperceptible.

> There is no longer an economy. Rather, there is *ecotechnics*, the global structuration of the world as the reticulated space of an essentially capitalist, globalist, and monopolistic organization that is monopolizing the world. The more the monopolization of the world makes the phantom of another "economy" disappear ... the more clearly *ecotechnics* displays what is henceforth possible: either ecotechnics takes on the sense of the autism of a "great monad" in a process of indefinite self-expansion, and/or it takes on the sense of the disruption of all closures of signification, a disruption that them up to the coming of (necessarily unprecedented) sense. That is, either ecotechnics is the entire sense of labor—of a labor henceforth infinite, dazed by its own infinitude and by its indefinitely growing totalization—or else ecotechnics opens labor up to sense, in-operates labor unto the infinity of sense. (Nancy 1997, 102)

6 The Racket

There is nothing astonishing about the fact that the "crisis of sense" is, first of all and most visibly, a crisis of and in "democracy."

—Jean-Luc Nancy (1997, 165)

Previously I have argued that John Cage essentially had to adopt the position of the cyborg in order to hear sounds in themselves and that this pattern has been repeated via the use of audio technology by sound artists and theorists, who, I argued, developed "electronic ears" in order to make the inaudible audible.[1] Here I want to look at the environment required to hear sounds in themselves: the silence that needs to be imposed and the relationship that silence has to present political, physical, and social configurations. Let us return to this moment, in 1952, when Cage entered an anechoic chamber and for the first time heard the humming of his circulatory system. It's difficult to underestimate the importance of this event in cementing a path Cage had been preparing for some time—whether consciously or not. The sound of the inner workings of his body, audible as a result of the silence of everything else, provided material proof that sound is everywhere—even in silence—that the hum of life is noise, and that noise is non-intentional, since, as he remarked, "no one means to circulate their blood." It was this experience that led him to compose *4'33"*—his famous silent piece, during which the virtuosic pianist David Tudor sat at the piano without playing a note. What the audience heard instead were the sounds of everyday life: a train rattling in the distance, birds chirping, and, depending on the recording, voices in the background. Whereas *4'33"* demonstrated that there was no such thing as silence in the external world, and indeed the concept itself is an artificial abstraction, the anechoic chamber experience demonstrated—for Cage at least—that noise exists as an essential part of (human) being, and its continuous hum penetrates the universe of sound, even when it can't be physically heard. From this experience,

Cage not only critiqued the use of silence in composition but also declared that there was no such thing as silence—the concept itself was an aberration. What filled the void of silence was the sound of the world, the sound of everyday life, the flat democracy of all sounds taking their place in partnership with life and art.

Quiet the piano, eradicate Western music, and ambient sound emerges. Quiet the room, eradicate the possibility for ambient sound, and the noise of the body emerges. The sound of ambience eradicates the concept of silence. The sound of the body eradicates the possibility of nonexistence, as its autonomic nature combines the hum of life with non-intentionality. This hum ensures that there will be a future for both the listener and the institution of music in general. In the present context, the continuity of media sound that flows into one's ears as an almost unintentional or "natural" event suggests something that will keep on continuing, ensuring a future for the individual and the modern subject in general. But what are the conditions, both physical and existential, for this continuation? For the "silence" of the body, the "silence" that guarantees its opposite, develops an existential and one could almost say logical paradox: not hearing it is impossible, but yet it can only be heard through isolation and amplification. The hum of life is as natural, as accidental, and as "ambient" as the circulation of the blood, yet to describe it as something that is "heard" seems awkward: just as "no one means to circulate the blood," surely also no one intends to hear their blood circulating? To this extent, could we call the sound of the blood something that is *heard* at all?

The Cagean listener requires the anechoic as much as the anechoic forms the Cagean listener. For what happens to sensation in the anechoic chamber—in the space where the individual joins with the world and existence through an architecture of silence? After his experience, Cage wanted to put miniature anechoic chambers over everyday objects in order to hear their internal vibrations. Some sixty years later, Cage's wish has been partially granted as the sonic habitat of the contemporary Westerner is one that physically reproduces, on a small scale, the experience of anechoic chamber. Urbanization has produced living environments designed to dampen the external world and attenuate the reproducibility of the sound within—whether it be voice or media sound. Acoustic insulation provides an environment for the clear transmission of human and media voices, while the incorporation of multiple sites for electronic communication has produced an acoustic habitat especially designed for hearing small sounds: the beep of the cell phone, the important news coverage occurring in the next room, the single sound of an e-mail notification. Yet, despite its discursive diversity, the aural range of this environment is very limited. The requirements of transmission demand a certain frequency range, and the ubiquity of mobile devices narrows this range even further. This creates a sonic homogeneity that enables a very fine-tuned hearing, only by disabling sound from the outside. The sound heard is further diminished by being aural only: this might seem tautological, but as mentioned, sound as vibrating energy, carried by air and atmosphere, is a phenomenon both heard and felt. Reduce the atmosphere, create a "dry" space, and the experience of resonance is necessarily truncated. The "dry" atmosphere, approaching the "dead" space of the recording studio, is filled with similarly dry voices, recorded in locations chosen for their lack of ambient sound, their freedom from the interruptions of noise. These dry and somewhat toneless voices bounce back and forth within the insulated walls of the average home, creating the kind of environment where one hears one's culture in much the same way that Cage heard the inner workings of his body, as an environment. The noise/sound of his body, routed through the microphone, amplifies not just the inner workings of his various internal systems but the whole concept of non-intentionality, a concept that, we find, requires the erasure of the very materiality necessary for sound to resonate.[2]

It is only really in the last half-century that, à la Cage, we have created an environment from our inner workings, that we have sonified a form of solipsism. The anechoic chamber seems at first to be an extreme instance or manifestation or materialization of this cocoon, now concretized in consumer-driven, networked individualism. But this is not an anechoica that we notice because wireless, and the media in general, often swamp all other sound, creating a pressure that forms its own habitat. Mobile media might move us out of the house and into the world, but the world is now domed by a data cloud. This habitat is now our sensible, extended from house to city and very soon to a world blanketed by networks. The energy of wireless sensation, replacing the low frequency waves of the wind and the sea, forms what has been referred to as a second nature, and popular discourse over the past decade has vociferously argued that this second nature is equal to and has a correspondence with the first, such that there can be some kind of dialogue, conversation, or even negotiation between the two.[3]

Indeed, in the most rarefied spheres of discourse there is no difference between the pockets of air that circulate abstract speculation and the gusts of air that rustle the leaves and vibrate the skin. There is no difference between the whispers of data circulating the globe, processing information, yielding statistics, activating filters, triggering transfers of money or opinion, and the murmurs, sighs, and gasps that propel a group's movement with an unshakeable conviction—if only for a moment. The world is now flat, we are told, and assigns no special privilege to the people beside you, breathing the same air, sharing the same neighborhood, state, nation, climate, or geography. Voices are flattened, condensed through transmission, while tonality—that quality unearthing multiple meanings—is lost in the crowded, placeless, speed of communication, or crushed by the overblown acoustic markers of affect, leaving listeners bemused, uncertain, and questioning their hearing and their judgment.

The Sound of Spin

It has been suggested that the sounds Cage heard were not the humming of his circulatory or nervous system, but in fact tinnitus. Tinnitus, from the Latin *tinnier* (to ring, tinkle), is defined as the perception of sound or noise that has no external source. So I wonder, could Cage have mistaken the sound of life for the symptoms of a noise-induced deafness? And when he announced after his experience that "until I die there will be sounds. And they will continue following my death. One need not fear about the future of music" (Cage 1961), could he have mistaken music's future for ours—specifically, the future of our listening, as it shrinks to fit various enclosures and substitutes the high-pitched whistling of rhetoric for the movement of wind and sea. Tinnitus is also associated with the jingling of coins in monetary transactions. As they say, America turns on a dime, and here jingling and "spin" converge, creating the impression of movement and therefore sensation—as if dialogue is going somewhere, as if debate is escalating to a point where something or some action may occur. "Spin" substitutes for actual physical sensation, and in its substitution creates a habitat, an ersatz environment, from ubiquitous screens that broadcast the constantly shifting economic and political landscape, in a continuous, vociferous stream. Spinning is often also associated with vertigo and tinnitus—with situations where the internal mechanisms are out of balance and the body becomes its own stranger—and in a certain sense, it does. In chapter 5, I mentioned the movement of tone and sound into language, via poetic speech, as walking a fine line between sense and nonsense. In the present discussion, it is important to emphasize the power inherent in this exercise, for which the expression "coining a phrase" is revelatory. As Agamben points out, the coin of the realm has, throughout history, enjoyed a special relation to signification, having been used as an exemplar for the special category of signs, "whose signator is the human being" (Agamben 2009, 38). Etymologically connected to the phrase "to coin" (Latin *signare*), there is an element of self-referentiality in coinage that epitomizes both the current global monetary system and the individual's relation to that system. Historically and theologically, the coin is a sign within which power resides both overtly and as an enigmatic, quixotic undertow—for the value imprinted on the coin is less significant than the promise—to honor, authenticate, and eventually repay the coin's worth—that its insignia represent. The exchange of coins is thus always an implicit contract with the monarch or the state to whom dues are owed.

This debt is twofold in Judeo-Christian culture: with Christendom, there is the injunction to "give unto Caesar" what is rightfully owed; but in giving, the coin, as sign, is also a token of original sin since the original humans, Adam and Eve, were unmarked before their fall from Eden. Given the status of emperors throughout the ages, it is no great leap to see in the markings of the coin or in the image of head of the state not just the obligation to pay taxes, but to repay the original debt inherited from the creation of the species itself (a debt which, as we will discuss, is impossible to pay). Compacted within the small and seemingly innocent coin, then, is a particular extraordinary relation, one which, as Agamben writes, "displaces and moves [the semiotic relation between *signans* and *signatum*] into another domain, thus positioning it in a new network of pragmatic and hermeneutic relations" (Agamben 2009, 40). This network becomes even more theologically driven by Bohme, for whom the signature's origins and power lies in the word of God. As we have seen, Biblical doctrine unites with sound (the spoken word) and music to influence the musical and cosmological theories of Kepler and Leibniz. The sign of the coin, one could say, is thus theologically inscribed:

> The aporias in the theory of the signature repeat those of the Trinity: just as God was able to conceive and give shape to all things by means of the Word alone, as both the model and the effective instrument of creation, the signature is what makes the mute signs of creation, in which it dwells, efficacious and expressive. (ibid., 43)

The aporias that Agamben refers to are relevant in the present, supposedly secular, era. Indeed, according to Agamben, secularization does not separate church and state so much as it "acts within the conceptual system of modernity as a signature, which refers it back to theology" (ibid., 77). Theories of the sign have been popularized in postmodernity with the Baudrillardian notion of the simulacra and simulation, in particular influencing popular culture (the film *The Matrix* is a perfect example) with a range of theoretical orientations that, often focused on media, have tended to diminish the profound repositioning that aporiatic thinking enables. While the laity in medieval times may have resorted to ritual, dogma, and faith in order to resolve the paradox of the Trinity, the modern citizen must resolve the paradox of a market that, with monetary policies like quantitative easing for instance, goes up *because* it goes down. Without a deity to arbitrate, the question of coinage—of who or what can vouch for the value of the coin—and indeed, the actual authenticity of the valuation, is constantly reopened.

In a recent talk, Michael Hardt (2012) mentions the ancient Greek tale of Theogenes's encounter with the oracle at Delphi, who instructed him to "falsify" or "change the face of the currency." Hardt argues that this is what has been happening in the financial crises, as immaterial, biopolitical production (such as "ideas, code, images, pharmaceuticals, affects, etc.") is brought into finance capital, and in the process, what was once considered priceless or common to all is given a value.[4] This process is, Hardt suggests, a form of "changing the face of the currency ... characterized by the dominance of rent as the primary mode of capitalist expropriation." The question of measure is paramount here: "Finance in general and derivatives in particular serve to stamp measure on, or translate into economic value, some productive processes that are fundamentally immeasurable." But, like the signature of the coin, changing the face of the currency also involves controlling valuation itself (Hardt 2012). This is a process of assigning not only monetary value but also all forms of societal value. Internalized by the individual, it demands a constant self-valuation based tautologically on a perceived regime of measurement that reproduces the power-driven, competitive, and increasingly fragile networks of finance. Thus positioned, the individual attempts to balance the contradictions inherent in being both a competitor and a citizen. Staggering, their fall is the fall of the markets, of capitalism itself, as the original guilt and debt that stands behind the currency asserts its authority while evermore concealing its face.

We can think of tinnitus and spin, then, as metaphoric side effects of "the racket"—which is at once a scam, often involving corruption and extortion, and, like the big bang, the sound of its own making. The racket names, through sound, an operation that seems to lack a before and after, that touches on the infinite and the archaic, that is represented by the mundane but universal existence of currency, and circulated by coinage endowed with a debt obligation and weighted with a guilt that seems as old as Adam. But unlike the deep tones of liturgical chant, and despite its historic gravitas, the racket has its own, peculiarly light, sound and timbre, and gathers in its circulations various sonic descriptors—spin, gibberish, buzz, chatter—that join with terms such as "bubble," "jittery," and "frothy" to indicate the high-frequency zeitgeist of contemporary political and financial discourse. There is much that these descriptors reveal; however, one aspect needs to be emphasized. As I have suggested, recourse to onomatopoeic terms signals a breach, or aporia, in thinking, as if sound (or tone) can bridge the contradictions that intellect refuses, or as if the echo or fricative ring of their articulation will prolong or focus thinking to produce an explanation. In this sense, sound facilitates a mode of thought less hampered by the prescripts of sense and, as I will discuss, offers one way of negotiating the conceptual blockage that secularized rationalism presents.

Financial Noise

The racket, then, is both acoustic and financial, representing bad sound and bad, often illegal, financial practices. In its financial sense, the racket trades on the noise of unmeaning: the lack of transparency, the indetermination, the unexplainable volatility of the market; the financial "noise" associated with statistical abnormalities, probabilities, and synthetic derivatives; and the epistemic opacity of financial instruments that are beyond the understanding of Nobel Prize–winning economists. In its acoustic sense, the racket is the noise of media "spin" generated by politicians and pundits, advertisers and experts, legislators, commentators, talk show hosts, and a long list of voices that emanate from an equally long list of screens. Unlike the sonic volume associated with machinery and industry, the quantitative volume associated with financial "noise" is linked to informatic power that is routed through systems and networks that have been engineered not so much to reduce the interference associated with signal noise but to extract meaning from its volumetric trends. The absence of sonic noise—the quiet on the trading room floor—seems to parallel, like a soundtrack, the course of capitalism—from an industry such as manufacturing, to finance as a major economic engine.[5] Yet its discursive traction highlights the need to take sonic metaphors seriously, to use as leverage, so to speak, in calling out the bankruptcy of systems that trade on the vacuity, the un-meaning that naming the aural often unfolds.

In tracking the soundscape of the stock market, Urs Stäheli shows how the constellation of the street, public space, vocal sound, and financial exchange reveal some of the most telling relationships between noise and power. Referring to Charles Mackay's *Extraordinary Popular Delusions and the Madness of Crowds*, published in 1841, Stäheli notes how the "incredible noise" of the original stock market (located on the public street) was considered so loud that it was relocated to the garden of a nearby hotel. Once semi-enclosed in a private space, the mix of business and pleasure was both amplified and in a sense diluted by the acoustic profile it created: as Stäheli points out, the noise was contained, but also concentrated, becoming "a point of attraction in its own right" (Stäheli 2003, 248). Enclosing noise architecturally is also a way of concentrating it socially, politically, and economically. Contained within four walls, this noise reverberates and reflects within itself, producing its own amplification and blurring its constitutive parts. As a result, the noise of financial speculation became less of a celebration than a cipher to be decoded.

According to Stäheli, the variety of sounds that make up the aural ambience of the stock exchange takes on an additional quality, becoming a "soundscape," an aural signifier of not just a particular place (the stock exchange building) but, more importantly, a particular activity. At the same time, an educated and individualized listening, one that can be used to discern the vital information, figures, numbers, and orders hidden within the exchange's broad spectrum of sound, produces the discerning ear of the professional speculator. In contrast to the noise on the street, where social differences were less pronounced, the enclosed exchange establishes noise as "noise" only for those who do not have the facility to hear its meaning (Stäheli 2003, 250). This faculty belongs to the trader, as does the esoteric information required to "play" the stock exchange. Hearing "noise" rather than information, hubbub rather than the vocal commands associated with trading, is in this case an aural experience of exclusion via financial illiteracy. At a certain point in the history of the stock exchange, this din—fascinating, concentrated, and rich in information though it was—became too much. Stäheli recounts the invention of the stock ticker by Calahan in 1901 in an attempt to reduce the noise of the telegraph boys as they jostled for the most recent quotes. In yet another irony, common to both the logic of noise and the efficacy of technological remediation, a new form of noise is produced. The "ticker," as it came to be known, actually ticked—so loudly in fact that it became a source of fascination in itself, attracting crowds—where it was supposed to dematerialize into the telegraphic ether. The attraction here is volume—both acoustic and informational. "The hammering of the ticker is directly determined by the market: if there is lots of business, the ticker noise increases—if there is no business, the ticker may even come to temporary halt" (Stäheli 2003, 252). Volume trading now had an acoustic index, a material signifier in the form of a mechanical noise. Higher volumes could mean many things—increased volatility (the markets are "jittery"), a sudden revelation (insider trading), a natural disaster, political upheaval, and so on.

Today the correlation between financial noise and acoustic noise, between financial volume and material volume, is so minute that an investigation into the physical location of the original stock market seems irrelevant. However, it is important to register another correlation—that between trading and regulation. For as the trading floor has become enclosed within a particular building, it has also been quarantined from legislative oversight and protected from scrutiny by digital barriers as trading and exchange take place within the electronic sphere. In this context, "noise" represents a far more dynamic (as opposed to mechanical) system of trading—one based on high-volume "financial noise," described by Stäheli as the trading of stocks at a level sufficient to register as a trend or uptick but not sufficiently coherent to produce anything like a "buy" or "sell" statement. Like the presence of noise in language or poetic speech, noise trading is "nontechnical" and ambiguous, and noise traders are thought to be irrational, basing their decisions on emotion rather than reason. Yet this also hides a myth—that there is such a thing as "normal" (non-noisy) markets. In the same way that extreme weather events are now within the "new normal" of global warming, extreme fluctuations in economic cycles are now part of a new dynamism entering the global economy, producing degrees of volatility of which the 2008 financial crisis was just one manifestation. Talk of "perfect storms" and "financial tsunamis" aligns the new reality of trading within familiar meteorological patterns, implying that a stable system prevails. However, in both the ecology and the economy there is now no regularity, in part, because there has been no regulation.

Engineering Bubbles

It just so happened that the research and development of REA, the real estate agent program discussed in chapter 4, coincided with the largest and most calamitous housing bubble in American history. As a prototype, REA may have at the time been seen as a commercially viable venture, offering an efficient, non-labor-intensive way to assist in the speculative real estate market. An autonomous agent making "robot sales" would bring a new meaning to the practice of signing mortgage documents without any fact checking, one of the key elements in the subprime mortgage meltdown.[6] Speculation aided by automation would bring the bricks-and-mortar real estate industry into a degree of compliance with the financial industry, which had marshalled the resources of newly minted mathematics graduates from MIT ("quants," as they were known) to create sophisticated and ultimately incomprehensible algorithms for the management of risk in the newly established derivatives market. The mythology of greedy agents and foolish homebuyers as the cause of the financial crises rarely includes the guiding hand of an algorithm or the agreeable image of an embodied conversational agent, yet it is within this orbit that the crises originated. Autonomous agents and robot sales describe an environment developing in the late 1990s that saw the rise of the dot-com bubble, the emergence of derivatives markets, and the subsequent explosion of debt (for, as Harvey points out in reference to the rise of fictitious financial capital, "Why invest in low-profit production when you can borrow in Japan at zero rate of interest and invest in London at 7% while hedging your bets on a possible deleterious shift in the yen-sterling exchange rate?") (Harvey 2011, 29). After the dot-com crash in 2000, Alan Greenspan reduced interest rates as a way of cushioning the economy from the downturn. Not only did this avoid a relatively "natural" correction in the market, it ensured the continuation of two mythologies: continual growth through house prices and continual progress through technology. The grand information age did not have to end—"clicks and mortar" and "bricks and mortar" would coincide in an ever-increasing inflationary bubble.

Low to zero interest rates were supported by an unbridled optimism in growth itself, in the potentials of self-actualization, in the productivity of technology, and in the endless possibilities offered by the new era of the virtual. With the plethora of new and exciting startups adding a silicon sheen to globalization, the need for regulation that would normally limit the risk of 0.25 percent interest rates was forgotten, ignored, and derided.[7] While the most hyperbolic strands of cyber-rhetoric faded with the dot-com crash of 2000, an enchantment with the virtual and the irresistible attraction of technological salvation nonetheless prepared the groundwork for the economic "virtualization" that was to follow, with the vaporware created by sometimes-nonexistent startups funded by venture capitalists and listed as IPOs morphing into the overpriced and underfinanced housing estates and apartment blocks that were marketed as homeowners' personal ATMs. Certainly house prices had risen for the past two decades, ameliorating the decline in real wages that had been steadily building since the '70s. However, this positivity in an otherwise declining set of living standards (e.g., lower real wages, longer working hours, inflation, higher costs for education) assumed the mantle of a natural law, so much so that the algorithms used to assess risk in mortgage applications did not factor in the possibility of house prices going down. This was not just an omission—there was no value or quantity expressed in algorithmic terms that represented negative growth. Here we have a rational set of numbers that omits the basis of its rationality; a mathematics expressed as an irrational desire rather than mathematics expressed or operating as itself; an economics based on the premise of "rational man," where the rationality has been excised from the equation; and the massive risk of indebtedness now managed by the "wondrous financial innovations of securitization that supposedly spread the risk around and even created the illusion that risk had disappeared" (Harvey 2011, 17).

This rhetoric of immateriality was substantiated by the entry of financial engineering into the markets. The accouterments of i-culture that impacted daily life (patterns of work and rest, productivity, and social relations) did so within a financial sector, which channeled massive investment into the research and development of financial instruments for the creation of derivative markets and mortgage-backed securities (securitization). During the 1990s, as Harvey points out, this "permitted a huge flow of excess liquidity into all facets of urbanization and built environment construction worldwide" (Harvey 2011, 85). At the same time, the catchcry of technological and organizational innovation produced its own market, which lobbied for increased deregulation while extolling the virtues of a market that, now part of the digital age, somehow always knew best.[8] Until recently, the effects of deregulation, globalization, and the risks of high-volume and automatic trading have been obscured by the mythology of "self-correcting" markets.[9] While a "correction" implies a return to stability after a period of either over- or under- valuation, the current financial crisis is more like the manifest signs of a "tipping point" rather than simply a "correction." This is because, as Harvey emphasizes, the only way to "correct" the crisis would be a return to 3 percent growth. Given the present economic and ecological limits to growth, such a return would be impossible without unimaginable consequences.

> For capital accumulation to return to 3% compound growth will require a new basis for profit-making and surplus absorption. The irrational way to do this in the past has ... been through war, the devaluation of assets, the degradation of productive capacity, abandonment and other forms of "creative destruction." ... Crises, we may conclude, are the irrational rationalizers of an irrational system. (Harvey 2011, 215)

The effects of this (ir)rationalization are felt in government-mandated austerity programs, collapsing real estate prices, obliterated pension funds, and general dispossession, orchestrated and administered by state and international institutions such as the IMF. The role of the state is essential here, since it is only through recourse to an external authority such as the Federal

Reserve—whose monetary policies of quantitative easing and low interest rates have upheld unsustainable growth—that faith in the markets and in capitalism itself can continue.

Many have wondered, myself included, why this massive redistribution of wealth has not produced more of a public outcry? The answer, in part, lies in the transformations imposed at the individual level. The "new economy" of the '90s and early 2000s produced speculative bubbles (especially in housing and finance) that drew upon speculation itself: as a reengineering of subjectivity based on consumerism, a heightened individualism, entrepreneurship, and hyper-optimism. These values may have once been attributable to "irrational exuberance" but now, in this era of sobriety, they have evolved into an overwhelming sense of guilt and dread. As Lazzarato writes: "With the bust of the new economy bubble ... the 'progressive' illusion which Silicon Valley, the dot-com economy, the new economy, etc., had implanted in people's minds, has given way to ... the power of capitalism to self-destruct of which the 2007 financial collapse was but one example."[10]

The end of capitalism, being unimaginable, easily blends with the prospect of planetary extinction as an inverse of the exuberance Greenspan warned against. For, despite the well-documented coup that Wall Street undertook (with the cooperation of the government), it is still the individual homeowner, shareholder, or pensioner who is targeted, both by the media and via government-funded schemes (such as those aimed at foreclosures) that, through endless and insurmountable bureaucratic demands, in fact reinforce the individual's sense of guilt and shame. Through whatever means at its disposal, "the party of Wall Street" (Harvey 2011) reshapes public sentiment so that "moral hazard" is directed toward the general population and away from political, ideological, and capitalist institutions.

In the predominantly Judeo-Christian cultures of North America and Europe, eternal debt and existential guilt have already been interiorized as "original sin" and can conveniently be combined with, or obfuscated by, environmental guilt. This debt/guilt axis opens the way for the individual, already defined as a consumer during the phases of debt accumulation, to be chastised for believing the rhetoric of continuous growth, for consuming too much, for allowing him- or herself to be "in debt," and, as a logical consequence, for having set foot upon the earth at all![11] As Christian Marazzi and others have argued, debt now subtends and controls finance on a global scale, and yet the character of debt, its force and its consequences, is targeted at the individual, both present and future.[12] This has led to the development and likely continuation of "the debt economy," such that debt "represents the economic and subjective engine of the modern-day economy," one that, while ordered via algorithmic processes, controlled by state institutions and regulated via the labyrinthine system that constitutes the market, is in fact based on the "moral" principles of debt and obligation that rationalize and legitimate the curtailment of public services, the increasing inequalities in wealth, and the appropriation of the common. Like Serres's concept of the sensible and Nancy's concept of the world, the common includes not just land but also all social relations and creative wealth. As Marazzi writes:

> The hidden face of financialization, of the recurrent production of "debt traps," as happened with the subprime bubble, is constituted by the production and exportation—silent, but real—of what we call the common. The common is the entire knowledge, understanding, information, images, affects and social relations that are strategically subject to the production of goods. (Marazzi 2011, 118)

Debt and the Indebted

> The figure of "indebted man" cuts across the whole of society and calls for new solidarity and new cooperation. We must also take into account how it pervades "nature and culture," since neoliberalism has run up our debt to the planet as well as to ourselves as living beings.
>
> —Christian Marazzi (2011, 151)

It would be impossible in the space of this book to address the various different financial mechanisms for transferring wealth from the public to the private or corporate sector, and to investigate the role of governments and so-called neutral institutions such as the rating agencies and the IMF. Such investigations are better left to more qualified authors.[13] My aim here is rather to discuss the ways in which these blatant transfers of wealth, together with the appropriation of public space and indeed the habitat itself, rely upon the individual absorption of indebtedness that has been occurring, in part, through the attack on "sense" and "sensing" in all its forms. Debt encompasses and absorbs the parameters of what we might call the global financial and ecological crises. Its perceived moral force overrides other social, environmental, and ethical considerations, placing all expenditure under the priority of repaying the debt, despite the fact that it is often not clear to those who advocate repayment just who the debtor and creditor are. Indeed, debt and its corollary guilt, whether environmental (climate change), social (rampant injustice), or economic (current and predicted crises), have become so internalized that the deathly overtones of "mortgage" (lit. "dead gage"; Partridge 1966, 416) are indissociable from the idea of "the future" in the sense that future generations will pay for the profligacy of those who will not live to experience its effects.

In his book *Debt: The First 5,000 Years*, David Graeber continually returns to the integration of moral norms, religious and cultural tradition, and the language of the marketplace. Asking "what precisely does it mean when justice is reduced to the language of the business deal? What does it mean when we reduce moral obligations to debts? What changes when one turns into the other? And how can we speak about them when our language has been so shaped by the market?" (Graeber 2012, 13), Graeber concludes that "to tell the history of debt, then, it is also necessary to reconstruct how the language of the marketplace has come to pervade every aspect of human life" (ibid., 89).[14] Yet while human life and the world it inhabits may have been financialized or monetized, the obverse doesn't hold—the discipline of economics is in fact founded on the exclusion of moral or "irrational" considerations. In questioning the common assumption that "surely one must pay one's debts," Graeber makes the obvious yet rarely mentioned point that the whole debt transaction involves a certain amount of risk, so if paying one's debts were a universal rule, if there was absolutely no risk involved, then the banks would fund absolutely anything (ibid., 3).[15] Risk is essential to the debt transaction, and in the latest financial crisis we have seen that risk is also something that can be financed, and the debt it incurs forgiven if the financier is "too big to fail." Risk is therefore a relative thing, as is the rule that everyone should pay his or her debts. Yet, as Graeber also notes, nowhere is this relativity voiced; rather, it has become a societal norm that one should pay one's debts: ethically, socially, as a good citizen. Budgeting to pay one's debts, hiring accountants to reduce one's debts, cutting social programs to retain AAA ratings in order to avoid accruing more debt—these are all "financial" practices, often based on minute calculation, that have at their base an accepted moral rather than an economic imperative. Yet the substance of this moral imperative is murky. As Graeber notes, "Consumer debt is the lifeblood of our economy. All modern nation-states are built on deficit spending. Debt has come to be the central issue of international politics. But nobody seems to know exactly what it is, or how to think about it. ... [Yet] the very flexibility of the concept ... is the basis of its power" (ibid., 4).[16]

As mentioned, the concept of debt and the contradictions and paradox it involves, are deeply tied to those of theology, especially Trinitarianism—most superficially in the figure of Christ, whose death "paid" for "our" sins, but more profoundly in the clash and the cognitive dissonance produced by any attempt to conceptually equate divine life and debt. Despite arguments that debt is the essence of society itself, that indeed there is a "primordial debt obligation" (what might be called an "original sin" in Christianity) beginning with the relationship between the individual and the divine, it is impossible to adjust or calculate the debtor-creditor relationship such that the debt can be repaid. [17] For as Graeber points out, "If our lives are on loan, who would actually wish to repay such a debt? To live in debt is to be guilty, incomplete, but completion can only mean annihilation" (ibid., 57). There is also the question of repaying the Infinite. Just as it was difficult to imagine the gods of Pythagorean cosmology not becoming bored with cosmic melodies, the course of which they always knew in advance, so too it is difficult to see how one can repay a debt when the gods are eternally and in every way complete.[18] There could be no logical or theological solution to this obvious imbalance between the finite and the Infinite. Where the figure of God or the divine is replaced by the universe—from which all life arises and to which an "existential" debt is owed—and from the universe, debt obligations filter down to ancestors, relatives, teachers, and so on, there is even less possibility of repayment because "if you cannot bargain with the gods because the gods already have everything, then you certainly can't bargain with the universe because the universe *is* everything. And that everything necessarily includes yourself" (ibid., 68). The problem then is not so much how to repay infinite debt or where to locate the beginning of debt or who might determine what this debt might be, but the concept of original, universal, and absolute debt in itself. One is born into debt in the same way that one is born into language or culture. However, isn't it more the case that one is born into the *concept* of debt and has to internalize that concept, take on that concept, and assume its structure in order for debt to work?[19] If the concept and actuality of debt are so baseless, why then does debt carry such a moral power?

The problem, argues Graeber, lies in the notion of exchange as the basis of sociality and, philosophically, the assumption of a prior separation that must occur in order for exchange to take place. For if it is difficult to determine a beginning of debt obligations, it is absurd to even attempt to determine an end. Everything can be subsumed under the rubric of the creditor-debtor relationship, from the smallest gesture—holding a door open—to the "priceless" sacrifice of a life to save another. In order to operate, this ever-present absurdity must be subdued and effectively ignored in everyday life so that one can take a certain amount of unremunerated collaboration, or neighborly goodwill, for granted. But what happens when the debt is, as in the current climate, infinite and unrepayable, who takes responsibility for repaying it? Why does the unimaginable size and imponderable nature of this relatively new debt economy immediately usher forth ancient feelings of individual and somehow original "guilt?" Could it be that the acceptance of the legitimacy of debt is itself a form of debt to oneself since it is a constant, gnawing reminder of the self's monetization within neoliberalism?

If accepting the legitimacy of the Trinity was an act of logical/mathematical cognitive cauterization, accepting that of debt is one of emotional self-abasement since it reduces all experience to exchange, where ultimately there is no arbiter other than the ephemeral market, and now, the nondiscursive planet. The question becomes not just "How much do I owe?" but pivotally, to whom, or what, can this question be addressed? The two sides of *oikonomia* that are now constantly under consideration—economy and ecology—are now equally indecipherable. "Answers" from the market—in the form of a rise or fall in various indices—are puzzling, especially when news of increasing debt ceilings, for instance, or continuing quantitative easing, produces an uptick in the market, or, during the bubble, when a cough or pause issuing from the throat of Greenspan could send the markets into a spin. "Responses" from the planet, in the form of rising temperatures and sea levels, are either the subject of intense debate or produce arcane models and simulations that, much like market indices, mark the disaster to come in percentages that mean little to the layperson.[20]

Drawing from the work of Lazzarato, we could describe such bewilderment as an existential condition, as "'the indebted man' at once responsible and guilty for his particular fate."[21] Because "the modern notion of 'economy' covers both economic production and the production of subjectivity. ... [the debt economy] also involves constant 'work on the self'" (Lazzarato 2011, 11). This encompasses the post-'70s emphasis on self-actualization, the pre-millennial rapture with self-virtualization, and now the post-millennial gloom of self-sacrifice arising from the implicit assumption that the individual is to blame for both financial and ecological debt. "Debt becomes a *debt of existence*, a debt of the existence of the subjects themselves" (ibid., 87). In this way debt questions the legitimacy of human life, subtending, and perhaps making redundant, the "debt to society" that is delivered at birth along with the "original sin" inherited from Adam. Like Graeber, Lazzarato sees this massive, universal, and infinite debt—the debt that can never be repaid—as an essentially theocratic force that structures time itself as a resource to be consumed.[22]

In this regard, debt and financial systems in general assume a power that was once given only to God. The medieval prohibition against usury was in part related to the acquisition of time that was only God's to sell or to own. Lazzarato cites a manuscript from the thirteenth century where usury is condemned as "the theft of light and eternal rest," where time is figured in days and nights. "The day [is] the time of clarity, and the night is the time for repose." If users "sell light and repose," then, this author continues, "it is, therefore, not just for them to receive the eternal light and eternal rest" (Lazzarato 2011, 48, n. 13). Imagining a divine ledger, this thirteenth-century writer argues that, having already appropriated eternal light and eternal rest here on Earth, it would be unjust for the stealers of time to also receive the same in the afterlife (ibid., 47, n. 12). The human rhythms of work and rest, where work is associated with clarity—knowledge, understanding—and rest with sustenance and recuperation, are as indissociable "as night follows day." But when debt usurps this rhythm, placing it within the "future-present" of a mortgaged future, the diurnal rhythms are disrupted—repose without rest, light without clarity, sensual experience without understanding. This association between diurnal patterns, lived experience, understanding, and sustenance is lost in today's economics, fundamentally, as the ethos of perpetual growth ignores the need for repose ("repose," also to rethink, a chance to reconsider, to "take a breather," and to literally reposition the self) and experientially, as the "light" is removed from daylight, "sense" from the senses. The presentation of economic data is almost entirely screen-based, the corporate office artificially lit and air-conditioned, the 24/7 "trading day" knows no time zones. As industry is given over to financialism and money is no longer associated with any material thing, it is as if time itself is in the process of being extracted from all sources of value, including the rest points and the pauses needed for growth to maintain its positivity. And just as the light-filled screens, continuously illuminate stock markets around the globe, the media machine that interprets and dispenses this data to the public also never stops. A form of praise, if you like, an endless song of fluctuating indices, simulating "growth" and change but affirming the existence of the market while discounting (literally) that of the individual, the nation, and life itself.

IPO: Helen Grace

A recent work by Hong Kong/Australia-based writer and artist Helen Grace explores the processes and effects of the internalization of the "indebted," the "work on the self," and the usurpation of time discussed above. *IPO: Helen Grace* was conceived in 2008, prior to the fall of Lehman Brothers, and originally intended to be a process piece that quantified a range of emotional states on a daily basis over six months, representing their fluctuations via candlestick charts in much the same way that, for instance, the commodity futures exchange represents the highs and lows of trading. Grace has been interested in the visualization of finance and economic information for many years, and wanted to create a work that both internalized and incorporated "the biopolitical reality of contemporary life, where there is always a question of what's profitable and what is not."[23] The work is also a reflection on the current form of neoliberalism, of "floating ... [oneself] into the stream of risky capital flows ... at a time when everything and everyone is devalued" (Grace 2009, 16). Taking this mentality to extremes, the work is a troubling reflection on the "corporate mentality [applied to] corporeality itself. ... It's an absurdity—but in a sense it does underpin the very absurdity of what is called the economic reality which we know" (ibid.).

In September 2008, Grace found herself in the midst of the GFC and decided to continue the work until December of that year. The work was first shown in Hong Kong in 2009; in that exhibition, there were 108 small prints based on monthly registers of emotional states between April and December 2008, a sound piece and accompanying notational track, and a small video installation. For the Sydney exhibition, the twelve emotional indices for October 2008 were enlarged as oil paintings. The audio component is an algorithmically constructed "composition" (produced in collaboration with Hong Kong new media artist Henry Chu) that ascribes tones in the lower register (below middle C) of the piano, to the rise and fall of the Hang Sang, and tones in the upper register to the emotional indices. In general, the Hang Sang is audible only when it rises beyond a certain level and so effectively functions as a background drone.[24]

Reminiscent of Russian constructivism, the red backgrounds and the white and black candlestick columns painted on canvas appear at first glance as an exhibition from the early twentieth century. The large photographs harken back to different eras in photography but have been painted by Grace to form a disturbing painterly/photographic hybrid. The first, reminiscent of landscape painting, is of women in the rice fields of Japan captured in the full swing of planting rice, overlain with a barely perceptible black and white candlestick graph showing the CBOT Rough Rice Futures in U.S. dollars for the month of April. The second, looking like a still from an '80s sci-fi film, shows the trading floor of the Hang Sang, brightly lit in neon green, displaying rows upon rows of computer screens in a room that seems emptied of people, overlain with red columns representing the Hang Sang Index's turnover for the month of October;

and the third, a black-and-white photograph of traders on the old Chicago Mercantile Exchange floor, overlain and almost caged by a chart of "Sleep per Night" for the month of October. Walking along the many rows of painted graphs and photographs charting Grace's "sentiment," the viewer can see the fluctuations in "general anxiety," "state of happiness," "everyday dread and dark clouds gathering," "dread and despair," "filial sentiment," "sense of security and belonging," "this volatile body, tired and aching," "are you lonesome tonight?"—that act as titles and descriptions for both indices and artworks. Like Ikeda's *Datamatics*, Grace's *IPO* charts the Infinite, the priceless, the unquantifiable, through a diurnal practice of reflection on the day's emotional states and their corresponding valuations. This is an exercise in reflection rather than strict correspondence, since as Grace points out,

> there is no real correspondence between the emotional indices and the Hang Sang, because it's impossible to say where zero is and where the end is. [For both the stock market and the emotional states] there's no reference point, or the only the reference point within the unlimited potential of the market is the previous crash. ... But that's the fantasy of capital, and of progress. ... That there will always be this inflationary potential for growth ... and it's a notion of growth that belongs to industry rather than nature ... and this also is the fantasy of economics, that it's linked only to industrial or mathematical time rather than cyclical, biological time. (Grace 2009)

The month that the Sydney exhibition opened[25] (June 2012) also coincided with a strange period in global financial, political, and governmental movements. Financial commentators ran out of synonyms for "disaster" and watched with a strange detachment the continuity of the markets amid the collapse of the economy that supposedly sustained them—which is to say that the stock market had detached itself as an index of economic health. There was an overwhelming sense of resignation—that the Eurozone would break apart, that China—up until then the world's economic engine—would begin to slow and draw the entire world into a black hole of negative growth, a negativity that would not be met by bloodshed, or military coups, or massive earthquakes, or the overthrow of regimes, a negativity that would have no expression other than the huge downward spikes that represented economic catastrophe almost four years earlier but in 2012 didn't seem to represent anything, the candlesticks having been burned already, the spikes having been whittled down by their constant thrusting into the solidity of the market. This is the body that Grace is charting, the tired body, the anxious soul, missing family and friends, alone in a city that has become home for an artist "interested in the connections between aesthetics and economics" (Morgan 2009, 7).

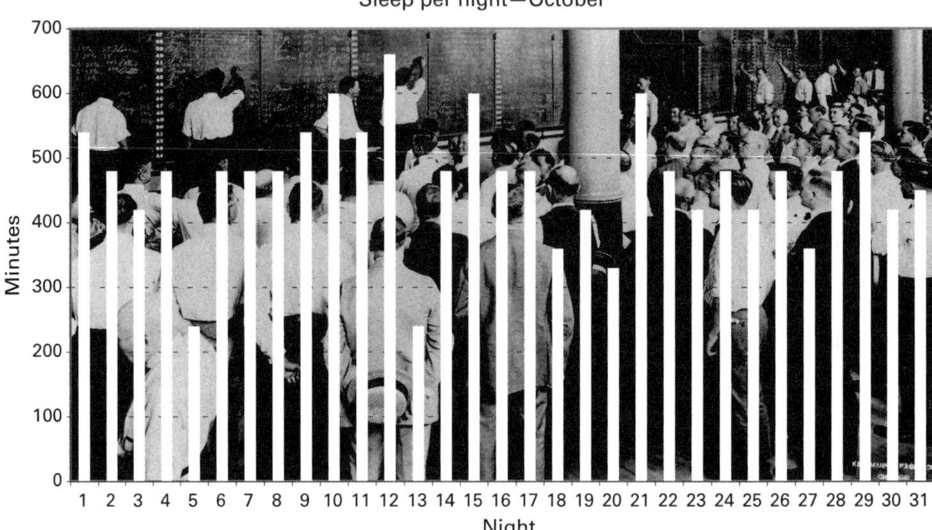

Figure 6.1
Helen Grace, "Sleep per Night—October," from *Speculation: The October Series* (2012). Oil on canvas. Collection of the artist.

In the catalog that accompanied the exhibition, Margaret Morgan suggests that Grace is in a sense "sending affects to the market" as an artistic response to the commodification and abstraction of subjectivity, as a kind of mirror image or a hyperbolic echo of what the market is in fact doing. "Quantify this!" "Put a value on that!" "You want 'market mood'? I'll give you mood!" As Morgan writes, "To insinuate the artist's most quotidian particulars into the stock exchange is to declare this informal value explicitly. We examine the data in search perhaps for some continuous and unchanged aspect of the self, and essential subjectivity" (Morgan 2009, 11). This is, after all, an artistic representation of a period of time that the artist lived through and, almost like a visual journal, is now depicted for all to see. "Helen Grace" is divided, abstracted, and exhibited. Up on the wall, "Helen Grace" is, as Morgan writes, perhaps, and this is also my interpretation, the "still life" that abstraction itself moves toward. "To transform anything into money is to abstract that form. To commodify flux, change ... is a distraction device to stop the inevitable, a way to stave off ends ... so that commodification, itself a kind of still life, wards off, by distancing, that actual ultimate end that comes to all things living: death." Morgan concludes her essay in the catalog suggesting that Helen Grace "deals in these impurities, these ghosts in the machine neither purging them nor running up their value, but rather opting for a kind of each-way bet: and the artist's value falls and rises on its strength and our trust." Grace's own description of this wager with herself, perhaps a necessary component of the "work on the self" that Lazzarato discusses, proceeds with the following:

I'm looking at my columns of figures & thinking, mmm, yesterday, on my *Everyday Dread and Dark Clouds Gathering* chart, I was a 32.19—so not too much despair there, and today, I had that great walk along Dragon's Back Ridge ... so I reckon I'm down to a 19.35 (because, you know, there's always some underlying or background noise, some hum of residual unease or apprehension that sits deep in the body, after this length of time living). Or maybe today, I had an encounter of cultural misunderstanding with a colleague and I felt homesick, so this has impacted on my *Are You Lonesome Tonight?* register, sending it to a new high. So this is how it went, hundreds of little decisions and judgments, in what I called my Decision-Based Emotion Capture System. There were 13 registers, each requiring four numbers (feeling at the beginning of the day, the end of the day, and the maximum and minimum levels reached in the emotional fluctuations of living) each day for 274 days—so 14,248 decisions or judgments in all. Now, you'd want to automate that if you were going to continue the system, but in the meantime, I was the recording machine, generating these numbers that would later form pictures—the Candlestick Charts that formed

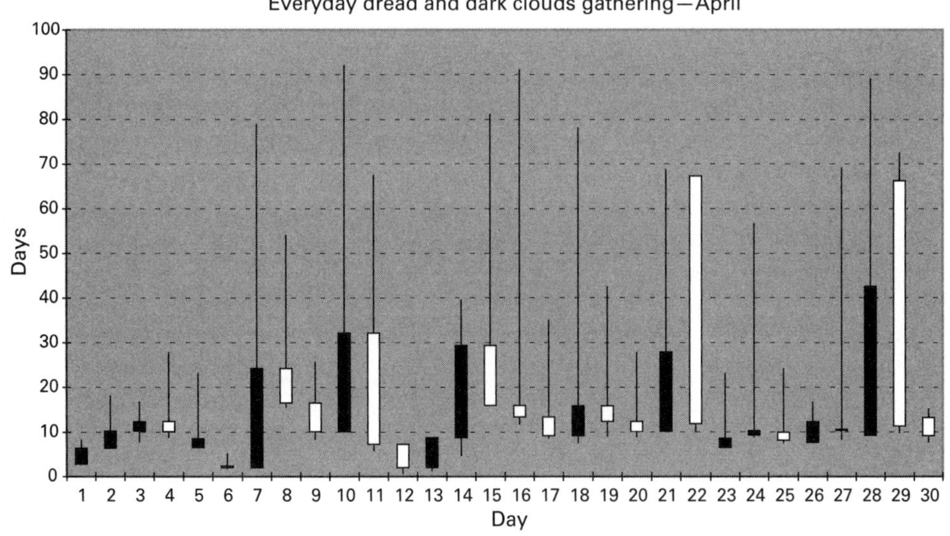

Figure 6.2
Helen Grace, "Everyday Dread and Dark Clouds Gathering," detail from *IPO: Emotional Economies* (2009). Inkjet print on archival paper. Collection of the artist.

the basis of the work—but I couldn't clearly see the pictures in the columns of numbers, I was only recording feelings, as clearly as I could. Part emotional reaction, part calculation—how to separate these two, in the moment of a decision or judgment? And it was important to be present to the process, rather than using pulse rates or galvanic skin responses or some such so-called "affective computing" process. This was, *above all*, affective computing, but I am using the word "computer" in the sense of its original human use—to refer to the women who did all the early calculations and entry in the development of the ENIAC for example, and are then somehow erased from the picture and we are left with an image of men "giving birth" to a *machine* called a computer.[26]

Grace's tongue-in-cheek reference to affective computing is in many ways appropriate. For the labor of distancing the self from the self in order to name, categorize, and quantify its affective states is not so far from the attempt to create a database from human vocal performances of affective states, as discussed in chapter 4. To some extent, it also corresponds to the schismatic role of the artist in contemporary culture, for the possibility of quantifying the infinite is as absurd as the possibility of having some object such as a painting that can both be sold and priceless at the same time.

In charting her daily emotional states, Grace is also making a statement about the still-current mythology of the artist as that person who can express (and quantify?), represent, and realize a public offering—that is, to speak as a representative of the public, to express the emotion of the times. The modernist references in her work—the atonality of the soundtrack, the Constructivist/Bauhaus look of the candle stick paintings—also reference certain artistic shifts that opened the way for a broader aesthetic, one that could, and would, eventually appropriate areas of life deemed beyond or outside of the proper domain of "art." For instance, while the twelve-tone system and serialism questioned the myth of the romantic composer, it also generated musical composition that enabled noise to be aestheticized and in a way contributed to an expansion of noise in other areas. According to Martin Scherzinger, "the link between Boulezian serialist practice and late capitalism … announced the future we are now living, as (apparently) unrelated innovations in the modes of control and domination. Boulezian serialism is the musical laboratory for this *mode d'emploi* for corporate cost reduction and its propaganda" (Scherzinger 2010, 125). As mentioned, the cultural industries helped develop the conceptual and artistic infrastructure for the financialization of everyday existence: body mapping, remote sensing, and mobile media contribute to software development and marketing; financial ties with Silicon Valley cement funding for the arts within the corporate sphere; and the discourse of posthumanism popularizes the algorithm as an almost genetic or biological construct within arts and culture.

As Grace points out, these are engineering rather than artistic innovations.[27] While posthumanism purportedly offers a critique of humanism, atonality a critique of Romanticism, and Constructivism a critique of expressionism, Grace reminds us of their historical and economic contexts: "What connects these reactions to Romanticism in the historical avant-garde of the 1920s and '30s is the Great Depression, so there is a whole sphere of economics underpinning it, and that is what then brings the artist back into that, more than the other way round." In the wake of continuing austerity measures, the artist, having "worked" on the self, being used to not just administering her time and energy but putting a value on her supposedly priceless creativity, might well consider selling her "stock" on the open market.[28]

IPO demonstrates that the terms of this "work on the self" can be mapped and charted, that an interior dialogue can, using the right techniques, be translated into market data, that feelings, though fleeting, can yield long-term averages, showing that Grace performed well in, say, difficult circumstances. But what happens when circumstances, such as the GFC, overtake the individual? "Everyone is their own boss" carries a continuous guilt regarding one's performance as a small business person, and this guilt is extended when one considers his or her ecological footprint (Lazzarato 2011, 134). Where the self is an IPO, and its "credit" or "stock" has been exhausted, shorting or even liquidating the self might seem the best option, just as the poverty inflicted by financial crises might, by putting a break on consumption, suggest one solution to planetary collapse.[29] This is to say that the endpoint of such ideologies, like the theology of old, always requires self-sacrifice, perhaps even sacrificing the self. But, as Graeber emphasizes, this too is absurd, for how can one repay a societal debt with one's life? And how can that life be measured in order to repay what is an unimaginable, unthinkable, infinite debt?

Conclusion: Echoes of Eco

As a way of highlighting some of the themes and questions raised in this book, let me return to sound poetry, experimental music, and sound art, considered in the broader network of technology, economy, space, sense, and resistance within which Stewart's voice-that-is-not-a-voice, resounds. First, recall that in her noise works, Stewart eviscerates the voice in order to relieve it (her) of the burden of subjectivity; second, this de-centered subject forms part of an ensemble that is noninstitutionalized; third, the democratic ethos that Stewart and other improvisers uphold is in part realized via computer technology, which provides a focus and an organizing principle both in terms of practice and (sonic) material; fourth, the collective output of this practice often requires a reorientation of the performer and listener towards a stillness and away from their usual movements and gestures; fifth, this occurs in environments that are outside of the gallery, institutionally, economically, and sonically—in that the venues are both uninsulated and often located within flight paths, the sound of which momentarily drowns out any autonomous sonic production. Finally, and perhaps paradoxically, this scenario encourages and perhaps even activates a listening practice that is unique in its ability to "hear" noise, within its environment and as an aesthetic experience that is shared and inclusive of all sound.

The corporeal self-consciousness and sonic self-annihilation I mentioned with respect to Stewart's "voice-that-is-not-a-voice" are also characteristics shared by the audience, as they quiet their bodies as much as possible within the highly reverberative spaces where many of the performances are held. If someone tips over a glass, its sound is absorbed as an impromptu and improvisational element, yet it is also accounted for, taken into account and made parenthetical to the performance. Sometimes the roar of planes overhead makes any absorption impossible: noise sounds as noise, an irritant rather than a deep rumble, a reminder of the miserable status of (sound) artists—their role being to physically occupy the unwanted, derelict, and sonically polluted parts of the city until such areas are ripe for gentrification, and perhaps sonically, to create an "adaptive" ambience such that the torture of low-flying planes and industrial racket can *almost* be heard as "art." Participants are decentered but also silenced; active contributors to the "work" but also subject to its self-disciplining demands; experimenters in the relatively new art form noise-music, but betrayed, or confounded by the need to exclude some kinds of noise (that which is produced by low-flying aircraft, for instance), and the desire to include "all sound"—as a category—within the performance. Stewart also wants to remove the subject associated with vocal sound but is unable to adopt the "nonsubject" of computer music because of the limitations of (her) physicality. This milieu is governed by an economy of noise within which sound, music, body, and voice are disciplined. Stewart's sucking sounds and her extreme efforts at "harmonizing" with the often computer-generated glitches of noise improvisers are an aural metaphor for a self-imposed absorption—an attempt to literally inhale the self in order to produce the noise of industry—the noise that the debt economy trades in.

In the expanse between an understanding of noise as a necessary byproduct of industry and a mark of difference, there lies not only a history but a gradual normalization and internalization of noise in much the same way that birdcalls in a particular area begin to mimic the industrial sounds of their environment. With Stewart's work, it is not so much the sound that is produced as the comportment that she needs to adopt in order to produce, say, the flat tones of anelectrical current flowing or interrupted sound. Her rigid and tense body, taut musculature, twitching finger, and "Frankenstein's monster" look give a visual face to this aural emission, as if, after so many decades of adjustments and finally acceptance of noise, we have to be reminded visually of its unsightliness. By holding her body in this way, Stewart draws attention to the kind of embodiment that the voice-that-is-no-voice produces. Similarly, the "democratic" impulse to accept all sound articulated by Cage and post-Cagean musics, to embrace noise as alterity, or as the sound of resistive forces on the fringe of Western art music has also tacitly advocated embracing the sonic dimensions of industry. After Cage, the glitch would come to symbolize the radical other in all its forms. Yet while welcoming "all sound" and eschewing all the traditional signs of music (for example, pitch, tone, rhythm, and timbre), proponents of noise rarely looked around to see what was happening in the environment from which and within which the sounds were occurring.[1] Could it also be that in repressing its presence and its disturbing effects, noise was shaped into something that *could* be absorbed along with the industry and the industrial-economic dimensions it belonged to?

This is more than simply a sensory adaptation. Return, for instance, to the beginnings of "noise-culture" in the 1960s and the great range of experimentation that was occurring in music and sound art in organizations such as Experiments in Art and Technology (E.A.T.). It is not insignificant that these organizations were also connected with the precursors of globalization and finance capital, just as they were also connected with the symbols of consumer culture (such as E.A.T.'s contributions to the Pepsi Pavilion at the Osaka World's Fair in 1970). Attali's *Noise* (Attali 1985) was published just seven years after E.A.T. moved its headquarters to Automation House—a move that may have seemed serendipitous at the time but in retrospect was deeply symbolic. The aim was to soften organized labor's resistance to so-called "labor-saving technologies in the wake of rising unemployment" (Harvey 2010, 15). In the process, the relationship between technology, automation, and the multitudes was rearticulated, rhetoricized, and rehoused (literally) in a building notable for its interior design. The move was publicized in the *New York Times* and praised in the design magazine *Interiors*, which published an article entitled "Automation House: Confronting Tomorrow's Problems behind Yesterday's Façade," lauding both the social policy and design features that Automation House embodied:

> The sponsors learned early that neither money handouts nor "make work" coped satisfactorily with the deep social malaise precipitated by automation. They saw the need for the participation of many specialized groups in communication, education, science, and art—to teach, demonstrate, and exhibit—and to capture the imagination of the public, *involve* the public. … E.A.T. commissioned the electrified, cybernetic, multimedia works of art, which will be so dominant in the interior. These works done by artists in collaboration with engineers, and achieved not with brushes and chisels but computer-age tools, are in no sense museum fixtures.[2]

These physical yet highly symbolic relocations and rearticulations correspond to a process of depoliticization, normalization, and eventually, as the present climate attests, subjugation. The same operation occurred—albeit on a much grander scale—in the 1990s, when globalization and the super-information highway (as it was known) became the new vehicle for a rearticulation of labor. If art, music, sound, and noise smoothed the way for the acceptance of automation in the 1970s—the emulation of artistic practice (such as the growth in creative industries, the adoption of innovation and experimentation as market benchmarks, the corporate sponsorship of "new media" arts, and so on) also muted the threat of automation within a newly globalized marketplace. Similar technical and corporate pressures are both absorbed by and shape artistic practice in contemporary sound art within an environment of increasing noisification, where the rehousing of arts venues is not just an economic practicality but assumes cultural and symbolic significance vis-à-vis the appropriation and corporatization of public space. This brings into focus the paradoxes associated with noise, space, and listening in contemporary urban culture.

The economy of noise embraces disciplines across all fields: ecological (in that noise actually drowns out the voices of other living beings just as it stifles the possibility for their appreciation), economic (in that "financial noise" has become an instrument of financialism), environmental (in that the noise we hear is often connected to the production of toxins and pollution), and psychic (in that to some degree, noise smothers thinking). The origins of "noise" (from *nausea*, and its association with the groaning of seasick passengers as they vomited overboard) remind us of its association with stench, which, like sound, is carried through the movement of the air, as are emissions of all kinds. The realization that our predominantly visualist, materialist culture ignores the immaterial, invisible, and ephemeral sphere at its peril, is palpable now, in the pressure that the atmosphere is exerting. Heard, and felt everywhere, the roar of rising rivers, the thud of mudslides, the screaming gales of hurricanes, and more recently, the shouts of protest echoing through the streets reframes noise within a sticky mass of sound, sense, air, and movement—a mass that is material, worldly, and inhabited. Through this mass, we hear the intensification of multiple dynamics, the accelerating beat of collapse. We hear "noise" sounding finitude.

Serres brings this noise within language itself, within the texture of writing and the grain of his thought, and in so doing he offers an approach to the praxis of thinking, writing, and sonic creativity that impinges directly upon the crises at hand. Like Nancy, Serres directs sound and listening to the heart of being but extends sound outward, so to speak, to fill the atmosphere with a resonance that is dynamic and duplicitous, a reverberative cycle that gains momentum rather than fading quietly into the background hum of modern existence. This noise demands attention, not for its sound but for its material and environmental residue. Its force is two-pronged: to return to a previous example—air-conditioners propagate by raising the outside temperature, more efficient noise insulation makes the process of increasing noise levels more tolerable. The environment becomes increasingly noisy as a result of urban densification, improvements in insulation, and an increase in the number of air-conditioners. But the residue of "atmospheric conditioning" is the return of the atmosphere itself in another form, as eventful, problematic, uncontrollable, and devastating weather. As Serres writes, "The mute world, the voiceless things once placed as decor surrounding the usual spectacles, all those things that never interested anyone, from now on thrust themselves brutally and without warning into our schemes and maneuvers" (Serres 2003, 3).

The hum of air-conditioning masks unwanted sounds in part by producing white noise. What happens in that masking? For it is not just sound that is hidden, nor is it the "all sound" that "white" usually designates.[3] When David Tudor described the limit of tone or "sound/image" as its total oscillation, producing white noise, he touched upon one aspect of the masking that air-conditioning exemplifies. White noise, being heard only as itself, approaches and perhaps encroaches upon the epistemic limit of tone, randomizing it, scattering it into the cosmos along with irrational ratios and big bangs. With the limit of tone breached, it is as if white noise is able to accumulate the modern version of cosmic and mythic power and to conceal in its ambivalence the industrial and symbolic transmission of power that it facilitates. In discourse, the limit of noise is also its oscillation between singularity and plurality. The difference between "a noise" and "a sound" and "Noise" and "Sound" provides a metaphysical tool for articulating a materiality that cannot be accounted for within the individuated world of objects or the aesthetic/communicative realm of tones. Like inside and outside the walls of the air-conditioned building, this difference designates a filter, what is in fact a noise filter that reindividuates through its mesh, allowing the passage from symbolic to worldly and vice versa. Previously I have designated this filter by capitalizing the "N" to emphasize the way that while Noise enables certain insights by blocking others, it doesn't necessarily recover the alterity that metaphysics and discourse require as an engine. In the current climate, this Noise allows us to ignore the other form of noise, that most directly linked to finitude—both the concept and the actuality. But what if Noise were treated as noise—banal, ubiquitous, and annoying? What if, following Serres's brutally frank descriptions in *Malfeasance*, noise is excremental?

> Inundated and deafened by advertising, who doesn't see an anus in the baffle of a loudspeaker. (Serres 2011, 42)

Serres might be referring to a form of coprophagia, but instead of eating, we are listening. The pollution we would never consider putting into our mouths is nonetheless blasted relentlessly into our ears. It is as if the echo chamber of media has switched registrations. Media noise has become toxic, disgusting—in a word, taboo.

The People's Microphone

Finitude is not a privation. There is perhaps no proposition it is more necessary to articulate today, to scrutinize and test in all ways.

—Jean-Luc Nancy (1997, 29)

In something of a reversal, it is now clear that debt is infinite, while the world is finite: both air and atmosphere, we are painfully discovering, have their limits, and the protective barrier they once provided is likely in the next two or three generations to become a shroud. The atmosphere, it seems, is also a container of sorts: a "home," a habitat that is not just being polluted by human activity but, like any enclosure, amplifies and transforms the subtle movements of systems within it. We are all familiar with the impact of human industry on the climate, and "eco" is often used as shorthand to specifically designate products and activities that have the ecology in mind. But this earth-oriented "eco" is preceded and modulated by its other common usage as "economy," and it is the latter sense that brings home the massive consequences of human echo-ing, the system of production and governance that has and continues to produce the economic and ecological disaster much of the globe is experiencing now, with the foreboding knowledge that it will only get worse.

The clang of metal in the forge was once the sound of industry, and the forge itself an integral component in the development of Western civilization. Etymologically associated with forgery, as counterfeit, and fabric, and fabrication, these days, its connotations are ignoble, and the raucous sound that once emanated from the forge's floor would now be associated with the equally egregious meaning of "racket." Like "forgery," "racket" seems to have lost any jovial associations with crowds on the street, carnivals, or festivals, and has become almost exclusively associated with various forms of illegal enterprises. It is no doubt coincidental that both "forgery" and "racket" describe the current state of capitalism dominated by financial markets. Both have developed over centuries, and both could be seen as the catastrophic manifestations of the tendency to ignore materiality and impose mathematical and abstract systems. With this, the materiality of bodies—the sounds that they make, the atmosphere they breathe, the ground on which they walk, the space they occupy, and the interactions that occur between them—is now under threat. As Grace remarks, "corporeality is being corporatized" with everyone becoming his or her own small business. The dominance of financialism, the calculations upon which its latest manifestation are based, the contradictions it contains, and the paradox of theopolitical governance that sustains it also threaten thought itself. Could it be that the mysteries of theism and the enigmas (to borrow from Harvey) of capital are designed to shape thought in much the same way that a stun gun shapes movement? This is to say that the paradoxes they spin insert a moment of paralysis, confusion, and disorientation in thinking, the return we might say of the Pythagorean comma within the cognitive operation. What fills the vacuum created by this pause, this momentary cessation in thought, logic, and speech? The incommensurable? Or is it rather a more mundane operative—one that administers and organizes knowledge such that certain groupings, hierarchies, and power structures cannot be questioned? Financial instruments such as credit default swaps and synthetic derivatives continue to go unregulated, in part because of their supposedly infinite complexity. The recent history of collusion between government and Wall Street remains largely untold, in part also because its implication—the end of democracy—is too disturbing. Alternatives to capitalism—well, they are simply unthinkable.

Debt implies the possibility of exchange, and for exchange to take place, value must be assigned, a system of measurement established, and an ontological decision reached that "this" is not "that," or that "this" equals "that." "Rational" thought, from *ratio,* always involves a calculation, a reckoning, a "rationalization," a judgment, and as such produces a thinking laced with all forms of separation and division. The elaborate system of exchange that forms the present economy has fully ingrained systems of valuation based on these decisions, not simply through decree (theocratic, political, economic, and "commonsensical") but as a *mode* of thought itself. By this I mean more than simply a way of thinking, a worldview, or a belief system, but a practice that has a texture, a grain, a tone, that leaves the thinker with a residue of sensation. Serres argues consistently that we must listen to the world: "Our voice smothered the world's. We must hear its voice. Let us open our ears" (Serres 2014, 42). This demands that we develop both ears that can *hear* the "voice" of the world, and a voice that can represent it, for "who will speak in the name of the silent partner whose worrisome rumbling is, little by little, covering the deafening noise of city centers and the booming sounds of the politico-media circus?" (ibid.).

As I have been suggesting throughout this book, the magnitude of the economic crisis and the corresponding poverty of thought and impoverishment of social relations that surrounds it calls for a radically different approach. "Crisis," however, is not the right word here since it implies a solution. Following Lazzarato, "catastrophe" may be more appropriate, yet it also implies a sudden event over which we have no control. Climate change, resource depletion, and the transfer of wealth have been going on for longer than a decade and require a collective memory that has, along with collectivity itself, been systematically degraded.[4] Because economic and ecological calamities are felt most poignantly and receive the most attention at the level of the individual and his or her immediate neighborhood, town, or city, it is almost as if the ongoing debt ceilings and defaults, austerity measures, political upheaval and corruption, or the extreme weather events, loss of agricultural land and habitat degradation are problems that can be solved, or need to be dealt with, at the individual level. But these crises cannot be "solved" by simply adjusting outputs, imposing regulations, or forcing austerity on households and their consumption, even though the myth that simply going green on an individual- or neighborhood-level is the most responsible course of action. "The indebted" that Lazzarato, Hardt, and Negri describe, lack the means to analyze the debt economy *within* this economy, and for this reason, as Serres cautions, it is time to cease endless rumination and speculation based on figures.

> The financial and stock market crisis that rocks us today is probably superficial, but it hides and reveals ruptures in time that go beyond the duration of history itself. … To access those buried causes requires that we leave today's financial data behind. … In other words, it is not enough to talk about the recent financial disaster, whose loudly proclaimed importance derives from the fact that money and the economy have seized all power, the media and governments. It would be better to accept the fact that all our institutions clearly and globally are experiencing a crisis going far beyond the scope of normal history. (Serres 2014, x, 17)

There are no words or models to allow comparisons between scales of the unimaginable. Just as medieval Europeans may have internalized the concept of infinite power—of God, Church, and King, to the extent that the comparisons of infinite power become senseless—so too there is a degree of pedantry in the presentation of scientific and economic data that graphs escalations: in the gap between the very rich and the very poor, the acceleration of species' extinction, the rise in ocean temperatures, the reduction in agricultural land, resource depletion, and so on over past decades. These graphs demonstrate a shift of such magnitude and consequence that the word "data" seems too simple, too fragile, too myopic to provide a foundation that could carry the weight and the gravity of the situation. "Data" can be contested, re-presented, negotiated in a way that leaves the magnitude of the global crisis held within the humanized sphere of imaginable and knowable facts, and as such, can be hidden within or obscured by discourse. The way that data targets, names, and quantizes the universe in Ikeda's *Datamatics* is a perfect example.

Graphs, figures, percentages, and predictions circle the globe as a layer of information that appears variously as chatter, information, warning, call to action, legislation, trading schemes, price rises. However, as this layer of part-knowledge and part-vapid speculation whispers across the surface of the world, the earth itself is shaken by typhoons and scorched by heat waves, presenting a voluminous response to the quiet conjectures of data. This is not a "response," per se—there is no dialogue here, only something noteworthy: that in terms of sense and sensing, the discourses available are mute in the face of the "natural" disasters that are now our climate. Who can hear climate deniers in a burning forest? Who can sign legislation when the lights have cut out? What this implies is a need to understand, to sense, and to form common sense differently—not through discourse or information, no matter how compelling, but through sensing: and this is where sound and listening play a pivotal role. Sound, as I have argued, offers a way to negotiate the "unthought" and the unspoken, to develop other vocabularies and other forms of political, economic, and social organization. Sound's ephemeral and atmospheric nature is, like the environment, something that circulates outside of exchange, and refocuses attention on the space and environment of the subject rather than the subject per se.[5] The aural opens avenues toward an understanding that is arational, that evokes a grain (or rather tone) of thought and an aesthetics of listening that, I would argue, offers some entry into the dilemma of how to hear the world and in hearing, also be able to act, with the aim and existential condition of the "in-common." From here, it might be possible to move toward a shared sensibility, a "communism of the senses" that builds sense, the common, and common sense simultaneously, for as Nancy writes: "All space of sense is common space (hence all space is common space). ... The political is the place of the in-common as such" (Nancy 1997, 89).

We can see this movement emerging in different though related cultural groups and practices. In terms of activism, the Occupy Wall Street (OWS) movement offered a complex and ongoing response to the privatization occurring on a scale unlike any other in Western history. Although taking perhaps two or three years to gain momentum, the occupy movement began a process of sounding and re-sounding the sonority that Wall Street, over the centuries, had managed to mute. OWS has matched the racket with an acoustic racket, amplified not through the corrupt channels of media but through the echoing of voices, transforming the crowd into an instrument: the "people's microphone."[6] Public space, the common, democracy itself, all values contained in the cry "We are the 99%" were expressed through sound, through many voices collected in a public space, opened to and often threatened by the weather, subject to physical abuse and state legislation, and corralled by laws regarding the use of public space, the kind of sound that could be made (no drumming for instance), and most of all, restrictions on amplification. This is an important and often overlooked point regarding one of the reasons why the occupy movement was prohibited from using amplification. It may not have been the sound or the loudness of the sound being produced, but rather the status that amplification gave to their speech. The voice, amplified, coming from a loudspeaker, directly confronts the role and monopoly of the media—to prevent the crowd from having a hearing and to contain noise as a disruptor in the ordering of sonic space itself. In this respect, the constant reference to the sounds of the crowd as "noise," "racket," and "raucous," the association of public speech in public arenas with incoherence, insofar as these voices don't cohere into any one particular "message") is not accidental.[7]

When related to the so-called Arab Spring, occupations in Spain, Syntagma Square in Greece, and riots in London that occurred in 2011, many commentators have noted the importance of corporeal occupation and a horizontal system of organization—such that there is no identifiable leader and everyone has a chance to speak.[8] "The people's microphone" acts as a poignant enactment of these features: it enables multiple voices to be heard, its echoes incorporate its participants—indeed the participants *are* the echoes, materially and consensually—and it presupposes the occupation of a common space where the entirety of the sensible and sensory environment is shared. Opinions, food, but also the weather, the atmosphere, the presence and absence of the police, the tracks in the ground, physical structures, etc., everything that marks out a physical environment becomes a trigger or an opportunity for a shared communication—even if that communication is only a raised hand or a blink of the eye. Hardt and Negri in their book *Declaration* remark on the almost mystical, certainly impossible to express or adequately describe, experience of being in the encampments.

> For professional politicians, and indeed for anyone who has not spent time in the encampments, it is difficult if not impossible to understand how much these constituent experiences are animated and permeated by flows of affects and indeed great joy. Physical proximity, of course, facilitates the common education of the affects, but also essential are the intense experiences of cooperation, the creation of mutual security in a situation of extreme vulnerability, and the collective deliberation and decision-making processes. The encampments are great factory for the production of social and democratic affects. (Hardt and Negri 2012)

Compare this communication with social media and the media in general, with the endless encouragement to "share," to vote, to "like," to answer. As an almost automatic response, the rights and wrongs of virtually every form of behavior and activity one might encounter are judged—even if that judgment is a click. Opinion then masks what is essentially a reduction in the conversational process. Rather than reaching consensus or debating an issue, the demand for some kind of moral judgment short-circuits the time it would take to reach a moral or ethical conclusion. From this, slogans are created, sides are drawn, discourse is truncated. In whose interests, we must ask? For judgment without consideration tends to produce platitudes and slogans that not only rely upon and enforce the ethics of capital, but reduce time and so the actual capacity for thought and ethical reflection. When Hardt and Negri conclude that "Facebook, Twitter, the Internet, and other kinds of communications mechanisms are useful, but nothing can replace the being together of bodies and corporeal communication that is the basis of collective political intelligence and action" (Hardt and Negri 2012), they suggest something that is known to all but has been denied since the ascendance of media technologies: that corporeal communication allows for different forms of both listening and sensing, for modalities that are far more sophisticated than any form of interactive, affective, or immersive platform. The twittering, babbling, echoing media and the murmurous crowd can be heard as the sonic materialization of a tone that supports a particular sonic/media ecology. But whereas echoing media perpetuate (literally) power-without-substance, echoing voices—on the street, in the market, at town hall gatherings, festivals, and demonstrations—articulate the common, public space in which sound resounds within and through a multiple rather than a unity. The repetitions in poetic speech and practices such as the people's microphone might be a form of echoing within language, a way of ensuring perhaps the duration or longevity of expression, but they are also a direct, acoustic imitation of reverberative space, which is a space populated by things, substances, people—a space of extension and a space of finitude. This physicality is absent from media-echo, as the resounding of sound, its resonance, occurs anechoically and anaerobically—in the absence of the environment in which echo can occur, and the air through which the reverberation is carried.

There is a general consensus that the occupy movement failed: it failed because it had no agenda, because it refused to participate in the democratic system, because it was far too idealistic in its demands, because, in a nutshell, it just wasn't organized. Some would argue that the movement's lack of coherence was also its strength in that it resisted incorporation into parliamentary-style democratic process with all its corruption and lack of representation. Two issues arise from the movement's so-called failure: first, we must ask, "Failure at what?" Second, if success is measured by incorporation, then we must ask, "Incorporation into what?" Recall in chapter 5 the double-edged sword of structurelessness—exemplified both in the "meaninglessness" of poetic speech and in the horizontal organization of movements like OWS—and Nancy's question: "What outline would retain the unexpectedness of sense ... without confounding it with an indeterminacy that lacks all consistency? What name could open up an access for the anonymity of being-in-common?" (Nancy 1997, 90). Incorporation into a coherent something: message, demand, spokesperson, agenda, organizational rules, and so on produces an identifiable voice, together with the amplification and distribution of that voice and its message. It is almost impossible to imagine a global movement of the size and ambition of OWS succeeding without some kind of media representative and/or representation. To say this, however, is also to suggest that such a movement is impossible "under the present circumstances." But does this count as failure? Where the impossibility of success is guaranteed in advance (with the prohibition on amplification perhaps being the first signal that the movement would never achieve the status of "representative"), does it make any sense to talk of failure? Perhaps there is a message, a vocabulary, a new approach in this "failure." For the echoes of the crowd faithfully repeating a speaker's message in a "profoundly Democratic" (Graeber 2013, 51) ritual, one that "discourages speechifying," create a sonic and auditory process or phenomenon that, like sound itself, fails incorporation. As such, it fails to answer the question of its own value as it fails to allow the possibility of an account. The point here is not so much to defend the movement against accusations that it didn't achieve anything substantial, but to suggest that the movement itself was and is a mode of expression that calls without calling, that articulates a need and an urgency without necessarily forming a demand, that presents a multitudinous vocalization running counter to what is regarded as normal and acceptable political speech. This vocalization is based on occupation—of an acoustic space that presupposes a collective and simultaneous embodied habitation.

What we might call resistive echo-ing, or echopraxia, adds another dimension, specifically an external, geographical dimension to the metaphor of resonance. To maintain this nuance, it may be helpful to make a temporary distinction between the metaphors of echo and resonance: whereas echo always introduces the environment (ecology), resonance can be an internal operation that also flows, logically almost, into culture; whereas echo is bound to death and earth, resonance multiplies with the promise of the immaterial and immortal. My focus on space is strategic—a way of moving past the association of sound with interiority and self-awareness (which also includes the notion of a self, an "I").[9] This association implies a scenario that has already established the anechoic. It has already established the law, the judicial system, religion, the possibility of absolution, and the space and protocol where this can occur. In this respect, Nancy's focus on the relationality involved in dialogue is important. Dialogue allows space, "the space between the replies," and this space is also fundamentally physical and communal (Nancy 1997, 165).[10] Like the improvisatory experimental music ensembles mentioned previously, the being-together, listening and responding to sound in a shared acoustic environment creates the conditions for dia-logue to flow in a sonic, interruptive, and perhaps instructive flow.[11] The movement of sound—both from the external to the internal and within the internal itself—creates multiple openings, which, for Nancy, "bears in the most extensive way the perceptible or sensitive condition as such." He describes this as a "double, quadruple, or sextuple opening," as a form of being both outside and inside, of being open from without and from within, of "sharing ... inside/outside, division and participation, de-connection and contagion" (Nancy 2007, 14). Sound returns to the emitter, rebounds, and plays within itself according to its characteristics: its volume, length, intensity, attack, partials, the effect of long-distance noises, and so on. Thus "sonority is not a place where the subject comes to make himself heard ... [but] is a place that becomes a subject insofar as sound resounds there" (ibid., 17). The echoes of corporeal communication rebind as they rebound, bringing together an experience that is qualitatively different from other experiences of commonality. Unlike music festivals, political rallies, and large-scale performance events, the voice that is being passed along passes through and is spoken by the participants, its delayed repetition inserting a pause into the conversational process. Unlike the contemporary paradigm of social interaction (the acoustically inhospitable café, for instance), there is an understanding that speech travels, and in this instance, it must travel from group to group. Speech, then, does not hover above each table in an isolated conversation bubble, is not whispered within a crowd during gaps in a performance, does not respond with an opinion or even a judgment, and therefore, importantly, is not obliged to grant the opinion maker a hearing—neither audience nor obedience nor the obligation to obey the order, injunction, prescription, advice, or call that is given.

The echoing speech moves outward, occupying the space rather than redoubling inward, producing either interiority within the subject or a monotone echo chamber within culture. Moving outward, the echoing group repeats without imposing a moral "answer" on the calls of another. Listening (*audire*, "to hear," etymologically associated with "to obey") is separated from obeying, and the latter from the obligation to enter into an exchange with the speaker. Adopting the people's microphone might seem a practice born of necessity; however, there is also a praxis that circumvents exchange implicit in its modality. For the concept of debt requires that a preexisting order has been established, for want of a better word, an "economy," one in which people and objects are brought into a system of exchange such that "debt" can be repaid. Repayment, however, assumes an equivalence, which as Graeber has pointed out is ultimately, logically impossible to establish. But more than this, equivalence requires isolation, separation, and unicity. It requires turning a phenomenon (such as sound) into a "thing" in order to set it apart from all else and give it a unique value. Just as Grace's *IPO* demonstrates the absurdity of attempting to quantify the emotions, other fleeting phenomena—such as the weather, the atmosphere, the tenor of social interactions, the goodwill of the community, or for that matter, long-term realities such as sustainability, generation, and even existence itself—are also beyond measurement. Without a proper place, dissolving into the atmosphere; subjectless: without the voice that would either presence or return (to) the self; identity-less: a mass that cannot be led and refuses representation—the sounds of music, art, and resistance may in fact be changing, if not the face of the currency, then certainly its sound. In doing so, they foreshadow both a withdrawal from the self-centered but also permanently indebted individual, and an occupation of space that is at once physical, cognitive, and sensual, which is "common" in all meanings (senses) of the word. Similarly, in basements, sometimes called "galleries," in lounge rooms or empty warehouses, people of all ages sit on old sofas or cushions on the ground just to listen to sound, or music, that is without analogy or reason, that seems at times to be brilliant and then again dreadful, in a way that is concentrated and focused, following the modulation of the tones, the nuances of the noise, with an intensity that seems, to me at least, driven by an urgent need to find a way to make sense, or re-sense, the crises that confront us. Regardless of the sound, the listening experience is shared within and through the atmosphere, almost as if the knowledge or understanding is parsed around the room along with the sound waves. The new vocabulary and conceptual tools for understanding indebtedness in all its forms might come from this sensory perspective, from this desire to sense differently, and to sense as Nancy would say “in-common.” Serres asks, “What if the present crisis in turn sounded the end of the economy’s exclusive reign?” (Serres 2014, 23). Traveling under the radar of rationales, evoking a sense beyond cents, summoning ancient and common wisdom—perhaps the end, after all, can only be sounded.

Notes

Introduction

1. The range of meanings attests to both the essential nature of all that is involved in both sensing and sense. For instance, from *Merriam-Webster's*:

> Sense: A general conscious awareness; a natural appreciation; the faculty through which the external world is apprehended; the meaning of a word or expression; what one must know in order to determine the reference of an expression; [ETYM: Latin *sensus*, from *sentire, sensum*, to perceive, to feel, from the same root as Eng. *send*; cf. Old High Germ. *sin* sense, mind, *sinnan* to go, to journey, German *sinnen* to meditate, to think: cf. French *sens*.]
>
> Sense: To perceive by the senses; to recognize. To feel; to become aware of; common sense. Sound practical judgment.

2. I should signal here that I am using "resonance" as it is popularly understood—not in terms of the acoustical properties involved in, say, sympathetic vibration, or resonant frequencies, in order to stress its sympathetic, empathetic qualities, qualities that amplify commonality. In this respect, I follow Nancy's emphasis on resounding, which multiplies, rather than simply redoubles, sound, in a return that brings to mind a history that precedes the sonic emission and is rewritten, or reshaped, by its return.

3. A more recent adaptation from the Greek root, and referring to the notion of habitation, and the relationship of organisms to their environment.

4. As both Agamben and Heller-Roazen point out, this dispute is not necessarily between singular and plural, unicity and multiplicity, monarchy and democracy, the one and the many, but rather, the possibility of forming a structure for the naming of being, a structure that would allow other systems, such as mathematical systems or musical systems, to develop without contradiction and that, as a bonus almost, would also complement the reigning theology that also faced the problem of accounting for the being of God.

5. This is not to deny that the system of notation that Western music has inherited has been vital to its development. The complex harmonies of, say, late classical and early romantic music attest to the necessity of very intricate forms of notation and harmonic figuration.

6. According to David Harvey, this amalgamation occurred in 1986. "The 'Big Bang,' as it was called at the time, linked London and New York and immediately thereafter all the world's major (and ultimately local) financial markets into one trading system. Thereafter, banks could operate freely across borders (by 2000 most of Mexico's banks were foreign-owned and HSBC was everywhere" (Harvey 2011, 20).

7. Available at http://www.evdh.net/portfolio/EvdH_portfolio.pdf.

8. Available at http://www.ryojiikeda.com/project/datamatics/.

9. According to Lazzarato: "The battles that once were fought over wages are now being fought over debt, and especially public debt, which represents a kind of socialized wage. Indeed, neoliberal austerity policies are concentrated in and fundamentally implemented through restrictions on all social rights (retirement, health care, unemployment, etc.), reductions in public services and employment, and wages for public workers—all for the purpose of constituting indebted man" (Lazzarato 2011, 127).

10. "Patrimonial individualism, whose basis is the assertion of individual rights, but according to a completely financial conception of these rights, right understood as securities. ... The beneficiary as 'debtor' is not expected to reimburse in actual money but rather in conduct, attitudes, ways of behaving, plans, subjective commitments, the time devoted to finding a job, the time used for conforming oneself to the criteria dictated by the market and business etc. Debt directly entails life discipline and a way of life that requires 'work on the self,' a permanent negotiation with oneself, a specific form of subjectivity: that of the indebted man" (Lazzarato 2011, 104, n. 7).

11. As Amy Goodman and her colleagues at *Democracy Now! The War and Peace Report* (KPFA radio, Berkeley, California) were fond of saying.

1 Endless Praise and Sweet Dissonance

1. Cited by Heller-Roazen (2011, 42).

2. Badiou continues: "The act of nomination, being a-specific, consumes itself, indicating nothing other than the unpresentable as such. In ontology, however, the unpresentable occurs within a presentative forcing which disposes it as the nothing from which everything proceeds. The consequence is that the name of the void is a pure *proper name*, which indicates itself, which does not bestow any index of

difference within what it refers to, and which auto-declares itself in the form of the multiple, despite there being *nothing* which is numbered by it" (Badiou 2007a, 59).

3. The term also becomes associated with rhetoric as the arrangement of material for an oration, distinguished from the concept of the sublime for its orderliness as opposed to a flash of insight. Agamben points out that contemporary theological scholars "speak of a 'traditional sense' of *oikonomia* in the language of Christianity (which is precisely that of 'divine design'), [and so] end up projecting onto the level of sense what is simply an extension of denotation to the theological field. ... In truth, there is no theological 'sense' of the term, but first of all a displacement of its denotation onto the theological field, which is progressively misunderstood and perceived as a new meaning" (Agamben 2011, 21).

4. Further, these harmonies could only be perceived by light, since "there are no sounds in the heavens" (Kepler 2002, 22).

5. Perhaps it is the sun—though how would we mere humans possibly know "what mind there is in the sun. ... Nonetheless, however things may be, this composition of the six primary spheres around the sun, cherishing it with their perpetual revolutions and as it were adoring it ... rings from me the following confession: not only does light go out from the sun into the whole world, as from the focus or eye of the world, as light and heat from the heart, as every movement from the king and mover, but conversely also by royal law these returns, so to speak of every lovely harmony are collected in the sun from every province in the world, nay, the forms of movements by twos flow together and are bound into one harmony by the work of some mind, and are as it were, coined money from silver and gold bullion" (Kepler 2002, 84).

6. "Accordingly, since we see that the universal harmonies of all six planets cannot take place by chance, especially in the case of the extreme movements, all of which we see concur in the universal harmonies ... and since much less can it happen by chance that all the pitches of the system of the octave ... by means of harmonic divisions are designated by the extreme planetary movements, but least of all that the very subtle business of the distinction of the Celestial consonances into two modes, the major and minor, should be the outcome of chance, without the special attention of the artisan: accordingly it follows that the creator, the source of all wisdom, the everlasting approver of order, the eternal and superexistent geyser of geometry and harmony, it follows, I say, that he, the artisan of the celestial movements himself, should have co-joined to the five regular solids the harmonic ratios arising from the regular plane figures, and after both classes should have formed one most perfect archetype of the heavens: in order that this archetype, as through the five regular solids the shapes of the spheres shine through on which the six planets are carried, so too through the consonances, which are generated from the plane figures, and deduced from them in book 3, the measures of the ex-eccentricities in

the single planets might be determined so as to proportion the movements of the planetary bodies" (Kepler 2002, 47).

7. "Just as in a chorus, when the leader gives the signal to begin, the whole chorus of men, or it may be of women, joins in the song, mingling a single studied harmony among different voices, some high and some low; for too it is with God that rules the whole world" (Aristotle, *Metaphysics, De mundo* [399a], cited in Agamben 2011, 72).

8. In many ways, the angels stand in for the public; however, Agamben points out that in the ecclesiastic tradition, the "politico-religious" aspect of worship, the "publicity," is "defined solely through the song of praise." Therefore, the liturgical aspect of worship resembles the song of praise of the angels and culminates in the Sanctus (Agamben 2011, 147).

2 Acclamation

1. Heller-Roazen continues: "which emerged from the currents that gave rise to the four preceding concepts, as well as from two traditions of diagrams: those of ancient Greek music, as related by Boethius, and those of the Neoplatonic and Pythagorean doctrines of the harmony of the celestial spheres" (Heller-Roazen 2011, 46). Heller-Roazen is drawing on work by Marie-Elisabeth Duchez (see Heller-Roazen 2011, 150, n. 16).

2. Durations also were given discrete identities, being based on a fixed and unchangeable unit of time—for instance, the *tactus*, which was a beat of moderately slow speed that "pervade[d] the music of this period like a uniform pulse" (Heller-Roazen 2011, 49, quoting Apel).

3. From a letter to Christian Goldbach, April 17, 1712 (see Heller-Roazen 2011, 84).

4. Leibniz also granted a role to incommensurable relations, giving them the ability to provide pleasure but only when in proximity to "rational relations," or only when they occur "accidentally." Descartes had considered the idea that musical intervals were in fact a particular rate of vibrations but concluded that this method would be too subtle "to be distinguished by the ear, without which it is impossible to judge the quality of any consonance, and when we judge them by reason, this reason must always presuppose the capacity of the ear." Heller-Roazen notes that for Descartes, only conscious representations could be admitted as a proper perception, and these sensations, multiple and minute, were too subtle to register (Heller-Roazen, 2011, 84).

5. "The hymn is the radical de-activation of signifying language, the word rendered completely enough with it and, nevertheless, retained as such in the form of liturgy" (Agamben 2011, 236).

6. Agamben adds: "That the Greek term for glory—*doxa*—is the same term that today designates public opinion is, from this standpoint, something more than a coincidence" (2011, 276).

7. Nancy has written on the relationship between Nazism, Wagner, and music. "It is not necessary to repeat what is well known about the powers and effects of music. ... It is necessary, rather, not to equivocate about the recognition of this fact: the resources of music [and of dance and architecture] in its capacities of harnessing, mobilization, and exultation were not invented by the Nazis. ... It is not excessive to state that this aspect of history belongs inseparably to religion and to music. Which means, by implication: to politics and to philosophy" (Nancy 2007, 50).

8. "In truth, resonance is at once listening to timbre and the timbre of listening" (Szendy 2008, 39).

9. Sterne uses the term "perceptual technics" to refer to the research, carried out especially in the 1920s by telephone companies such as AT&T, into perception underpinning the drive to economize signals in order to "monetize the gaps in human hearing" (Sterne 2012, 19).

10. Globalization was itself premised on the export of this culture, and the massive outsourcing of all bricks-and-mortar industries was seen as simply clearing the ground for higher-paid, value-added jobs and industries.

11. Hearing the end of sound, or rather hearing sound fulfill itself, moving forward, is one of the reasons why diabolic sound is often represented in film music as being back to front, literally played backward. This seems to be a reversal of any kind of natural order because it represents going back in time. Similarly, the truncation of sound in both sound art and experimental music through various techniques of manipulation, and in contemporary music through sampling, has its own signification as a manipulation of time, the future, and the listener's time and future. The future promised by the act of listening is, however, the very future that is in question. The manic emphasis on living the future, the general neophilia of consumer culture, borrowing from Nancy we might say the exultation of the future, is practiced on a knife edge of credibility, ever-threatened by the tipping point of the two major dynamical systems—ecology and economy—hurtling toward the West like category-five hurricanes.

12. In politics, however, where "just as the machine of the theological *oikonomia* can function only if it writes within its core a doxological threshold in which economic trinity and immanent trinity are ceaselessly and liturgically [that is, politically] in motion, each passing into the other, so the governmental apparatus functions because it has captured in its empty centre the inoperativity of the human essence" (Agamben 2011, 245).

13. So for instance, notable switches that Barack Obama makes between his Harvard and South Chicago accents—switches and variations that represent both African American authenticity and Harvard-educated respectability—have endeared him to the wider public since they unite in one voice such different and divergent communities. These variations in accent are, however, not as important as the amplification that his voice received through the media apparatus. As a presidential candidate, this amplification is essential, as essential as the voice itself, yet it is not strictly sonic; rather, it is mediamatic. We can think of it as a process that is applied to sound, that shapes it and directs it, and that is so integrated within the listening process that one could say it adds its own "timbre" or "tone." In many ways, one could think of it as a class accent: something so integral to speech that it cannot be extracted and scrutinized as a separate quality, but in the age of campaign financing, it can also be heard as the sound of money—a whirring, jingling, sharp-edged twang that folds itself around every announcement. Recently in Australia, politicians have taken to repeating every phrase twice, as if no longer relying upon the media to create the echo that is needed, but privately setting up a reverberation within the speech itself. This repetition, like the eternal singing of the song of praise, reaffirms, however mythologically, not only that someone is listening but that there is a being that reflects speech back. In everyday discourse, this would be another speaker or at least a body that reverberates. But there is no such body in the media sphere, only a collection of individuals forming a mythological "social body" whose unity is affirmed by opinion polls and elections.

14. Parts of this section have been excerpted from Dyson 2009b.

15. *Listening Post* part 1 is available at http://earstudio.com/2010/09/29/listening-post/.

16. In his book *Technoromanticism: Digital Narrative, Holism, and the Romance of the Real*, Richard Coyne identifies this modernized version of universal harmony as a form of "technoromanticism" in which, citing Csicsery-Ronay, "the computer represents the possibility of modeling everything that exists in the phenomenal world, of breaking down into information and then simulating perfectly in infinitely replicable form those processes that precybernetic humanity had held to be inklings of transcendence" (Coyne 1999, 73).

17. Philip Auslander describes *Listening Post* as a "performance" rather than an installation or an electronic sculpture, based on the argument that machines can be performers (Auslander 2005, 6). This argument follows Brenda Laurel's analogy between a computer program and a dramatic script that the computer executes—in a way similar to an actor. It might be objected that this view reduces actors to machines, whereas most actors interpret the text. However, Auslander claims that "interpretation is not essential to performance" and bases this assertion on the model of musical performance. Auslander follows Stan Godlovitch's distinction between "technical" and "interpretative skills"—a distinction that operates on the axis of "quantifiable" and "nonquantifiable" acoustic effects produced. Since the

computer in *Listening Post* produces the former, it possesses one of the skills required of a performer. The example Auslander gives to support this ascription is that of the orchestral musician who follows both the composer's score and the conductor's instructions. The similarity between the musician and the computer are sufficient to extend the thesis: the comparison suggests that "machines and human beings are equally capable of technical performance." The comparison is particularly compelling given that the performance of both is "live"—a facility that distinguishes the computer from a glorified playback device because they do not reproduce an existing performance: "the distinction is that between a technology of reproduction and a technology of production" (Auslander 2005, 7). Auslander means "live" not as an ontological category, but in the sense of being co-present physically (such as an installation, not a radio broadcast). It also has a degree of agency in its ability to sort and "frame" the multitude of messages and conversations occurring at any one time on the Internet: "*Listening Post* ... is a window or door that opens onto the world of the Internet and frames spontaneous activity found there as aesthetic performance" (ibid., 8). In this framing, the computer far surpasses the technical prowess of the orchestral musician. In short, the computer can be thought of as a performer and consequently "artistic performance is not an exclusively human activity" but belongs to machines.

18. Irun Cohen writes: "The difference between signal and noise, one might claim, is having the right reception. ... In informational terms, *Listening Post* is a machine that transforms noisy cyberspace information into a new narrative by selecting and recombining fragments of the flux. *Listening Post* dynamically self-organizes, similar to a living organism" (Cohen 2006, 1214).

19. Margot Bouman cites Adam Gopnik's review "Chatter: Orange and White" in the March 3, 2003, edition of the *New Yorker*, writing that *Listening Post* "is a machine that can tell you what the world is thinking—that actually listens to the world, reads its mind, and tells you exactly what's up in there" (Bouman 2003, 118).

20. "The layers of pitched voices take on the quality of a chant or liturgy as they blend with each other and with the reverberating tones. By the end of the [aural] scene, as many as forty voices can be heard and about two-thirds of the 110 displays show messages" (Rubin and Hansen 2002).

21. *Listening Post* part 2 is available at http://www.youtube.com/watch?v=QCwfw0v6mlo.

3 Infinite—Noise

1. As Kahn notes: "The fifth source was the world of unheard, inaudible, and 'supersensible' sounds, which pertains to the inaudible sounds permeating the cosmos. In the West these ubiquitous and persistent sounds were primarily the legacy of Pythagorean and Platonic discourses, but they also occurred as nondissipative

sounds and voices within a variety of mythological and fictional contexts, and they received added support during modernism by the physics and pseudoscience of atomic and molecular vibrations. ... During the late nineteenth century this Western cosmos resonated with the sympathetic vibrations of Hinduism, reintroducing Pythagorean thought as a whole by reconstituting its historical suppression of auditory space (as opposed to mathematical, proportional, intervallic, and musical space). As in private listening, these sounds could be heard only by select individuals or by individuals with the proper technology or promise of technology" (Kahn 2004, 114).

2. "What it gave out at first was a cawing noise, like a crow, a cynical old crow whom nothing fools or moves to pity but anything he finds defenseless for a moment, large and small, any living thing, he never lets go and finishes it off. Meanwhile I was making the instrument repeat and repeat again its desolating call" (Michaux 1992, 67).

3. Further, "Only because astrology (like alchemy, with which it is allied) had conjoined heaven and earth, the divine and the human, in a single subject of fate (in the work of Creation) was science able to unify within a new ego both science and experience, which hitherto had designated two distinct subjects" (Agamben 2007, 23).

4. According to Martin Scherzinger: "Arguably, the most ekphrastic deployment of music for philosophy in the twentieth century is the work of Gilles Deleuze and Félix Guattari. The second volume of their *Capitalisme et schizophrénie* is practically a study in inter-semiotic transposition, amalgamating the conceptual and sensual modalities (gestures, images, rhythms, sounds) of modernist music and those of philosophy." Scherzinger focuses on the way that Deleuze borrows heavily from Boulez, and the relationship Boulezian serialism has to late capitalism and, in particular, to the virtual economies of online media. "The link between Boulezian serialist practice and late capitalism, then, is to be found in this mystified process of desubjectification, in which mastery and domination are hidden; high modernist serialism would in this sense have announced the future we are now living, as (apparently) unrelated innovations in the modes of control and domination. Boulezian serialism is the musical laboratory for this now generalized regime of ... social control, a *mode d'emploi* for corporate cost reduction and its propaganda" (Scherzinger 2010, 125).

5. For instance: "In short, the Baroque universe witnesses the blurring of its melodic lines, but what it appears to lose it also regains in and through harmony. Confronted by the power of dissonance, it discovers a fluorescence of extraordinary accords, at a distance, that are resolved in a chosen world, even at the cost of damnation. ... In its turn harmony goes through a crisis that leads to a broadened chromatic scale, to an emancipation of dissonance or unresolved accords, accords not brought back to a tonality. The musical model is the most apt to make clear the rise of harmony in the Baroque, and then the dissipation of tonality in the neo-Baroque: from harmonic

closure to an opening on to a polytonality or, as Boulez will say, a 'polyphony of polyphonies'" (Deleuze 1993, 82). To add to the confusion, it is quite a jump to go from Baroque to neo-Baroque to contemporary classical or avant-garde music without passing through any of the other periods such as romantic or early modern. Although at the outset Deleuze has expanded the notion of the Baroque to include anything of a particular style, and therefore not limited it to a particular historical era, at the same time, it is very difficult not to suspect that he really does have the historical Baroque in mind and that when he uses words like dissonance, he is actually using them in a musical sense. This is one of the main problems with the use of the musical metaphor throughout because, discounting the caveats that it is only a model, we see an argument or a teleology that moves from melodic line (horizontal) to an interruption by harmony (vertical) to its expansion in terms of pitch, to chromaticism, polyphony, and presumably the microtonalities of contemporary classical music.

6. Sterne continues: "Far too often artists still fetishize noise as transgression or a challenge. Sampling, turntablism, mashups, and remixing all challenge the contemporary order of intellectual property, but they have not undermined it" (Sterne 2012, 125–126).

7. DeMarinis, in conversation with the author, San Francisco, 1990.

8. "Engineers began to imagine that they could move noise underneath other more desirable sounds. Noise could be masked and put in its place; it did not have to be eliminated" (Sterne 2012, 22, 94).

9. See Dyson 2004.

10. According to Sterne, Attali "adopted cybernetic language as a social theory precisely at the moment that communication engineering exhibited a new attitude toward noise" (Sterne 2012, 124).

11. Available at http://www.evdh.net/lsp/.

12. Available at http://www.ryojiikeda.com/project/datamatics/.

13. *Proportional Control System for the Festival of Art and Engineering* (196), The Daniel Langlois Foundation for Art, Science, and Technology, *9 Evenings: Theater and Engineering Fonds*, 9 EV B3-C6.

14. Interview with the artist, February 2011.

15. Interview with the artist, 2011. The "dandelion" is structural: moving between D, G, and the occasional B flat, the artist provides a basic harmonic structure from which overtones can be produced, pushing tonality to a level of complexity where it breaks up, microtonal complexes multiply, and give way to noise.

16. Incidentally, when amplified and compressed, this noise sounds like an explosion. The addition of rhythmic sounds turns it into a drone, which then becomes

the baseline for another movement—literally another iteration of the development of the complex dynamic system that the artist has created. Yet despite its complexity and its dynamism, despite the generation of theme and subtheme, despite the movement away from and then possibly recapitulation to primary tones, all sound returns to these two pitches—with the B-flat inserted as a discordant accidental.

17. See online documentation at http://www.ryojiikeda.com/archive/concerts/-dataphonics_concert_version.

18. Exhibited at *Event Space*, Carriage Works, Sydney, 2013.

4 Disaffected Voices

1. Parts of this chapter have been excerpted from Dyson 2009b.

2. As Timothy Lenoir surmised at the turn of the millennium, arguably during the peak of the debate: "These new media are shaping the channels of our experience, transforming our conceptions of the 'real,' redefining what we mean by 'community' and, some would maintain, what we mean by our 'selves'" (Lenoir 2000a, 298).

3. The use of the word "agent" is of course contested, and a full discussion of its implications is beyond the scope of this chapter. To avoid confusion, I am adopting the common definition of an agent as a software "entity" that can sense and respond to its environment and is autonomous to the extent that it is perceived as having an independent existence—even if that existence is confined to a computer screen, cell phone, wearable device, etc. See, for instance, Juha Vierinen: "An agent is an autonomous entity of software, which is able to work towards a goal by means of perceiving and communicating with its environment" (Vierinen 2002, 1).

4. Further: "The coming of the event is what we cannot and must not prevent, another name for the future itself" (Derrida and Stiegler 2002, 11). What does the event require in order for it to arrive *as* an event? For the event to be, as such, for the other to arrive, it must be invited, so to speak, without forethought, prediction, without the desire for assimilation or appropriation: "The child who comes remains unforeseeable, it speaks, all by itself, as the *origin* of another world, or at the *other* origin of this one" (ibid., 12).

5. Derrida adds that with these technologies "the border is no longer the border, images are coming and going through customs, the link between the political and the local, the topolitical, is as it were, dislocated" (Derrida and Stiegler 2002, 57).

6. "This other time, media time, gives rise above all to another distribution, to other spaces, rhythms, relays" (Derrida and Stiegler 2002, 7).

7. Artifactuality concerns the presentation of the present and its ability to drown out the other and must, according to Derrida, be met with "dissent," "dissonance," and "discord" (Derrida and Stiegler 2002, 9).

8. Cassell et al. explain these protocols: "Humans partake in an elaborate ritual when engaging and disengaging in conversations. ... Following [an] initial synchronization stage, or distance salutation, two people will approach one another, sealing their commitment to the conversation through a close salutation such as a handshake accompanied by a ritualistic verbal exchange. The greeting phase ends when the two participants reorient their bodies, moving away from face-on orientation to stand at an angle. Terminating a conversation similarly moves through stages, starting with non-verbal cues, such as orientation shifts or glances away and culminating in the verbal exchange of farewells and the breaking of mutual gaze" (Cassell et al. 2001, 57).

9. Cassell and Bickmore give the following example: Tim approaches REA. REA notices and looks toward him and smiles. Tim says, "Hello," and REA responds: "Hello, how can I help you?" with a hand wave. Tim says, "I'm looking to buy a place near MIT." REA glances up and away to keep the turn while "thinking." REA says: "I have a house" with a beat gesture to emphasize the new information "house." Tim interrupts by beginning to gesture. REA finishes the current utterance by saying "In Cambridge," and then she gives up the turn. REA finishes the house description and then continues (Cassell et al. 2001, 61).

10. See, e.g., Cahn 1990.

11. Note that researchers have relied on acoustic data to determine the emotional state of the speaker (the acoustic model) rather than relying on representations of the speaker's intentions, which describe primarily what the listener hears (Cahn 1988).

12. "If the user emphasizes a color, the Decision Module produces a speech act that involves commenting on the color ... if the user emphasizes an object ... the Decision Module produces a speech act which involves commenting on the object (Cassell et al. 2001, 63).

13. "The reality effect in no way guarantees the authenticity of what is captured. But it nonetheless remains the case that it elicits an authentification effect for the person who looks. Hence a certain mode of accumulation, in an 'exact' form producing a sense of exactitude and of authenticity, that is to say of presence, would be the condition of a certain form of intelligibility" (Derrida and Stiegler 2002, 108).

14. The full quote reads as follows: "The visual image plunges into a new form of knowledge because it knows that within its knowledge is inscribed an irreducible nonknowledge of the image" (Derrida and Stiegler 2002, 159). Stiegler later qualifies what he means by this "knowledge" as designating a "techno-intuitive knowledge" (ibid., 162).

15. In the same article, Cassell mentions the "GrandChair" project in this context, where a childlike ECA encourages grandparents to relate family stories.

16. For instance, researchers note that skin conductivity tends to climb when a piece of music is energizing and falls when it is calming (Picard 1997, 23–25).

17. Enhancing affective bandwidth through various devices is seen as a way to mitigate what Picard and Klein describe as a general societal lack of "solid, effective, non-judgmental active listening skills" (Picard and Klein 2002, 15) and has spawned projects such as Pentland and Eagle's "The Relationship Barometer: Mobile Phone Therapy for Couples," a project that would extract salient features—such as speaking rate, volume, duration, transitions, and interruptions from couples' conversations to determine which party is dominating the conversation, which party is doing the listening—and intervene accordingly. Collected over a period of time, the researchers believed that the results would not only yield information about the health of a couple's relationship but potentially intervene in the conversation by "strengthening the positive features while dampening aspects that have a high probability of propagating a negative interaction. ...[For instance] using such features as volume and pitch, automated mediation techniques [would] help transfer the floor to a participant who is being dominated in a conversation" (Eagle and Pentland 2003b, 1; see also Biever 2004). In addition, according to Pantic and Rothkranz, having computers monitor attitudinal states relieves humans from being present "to perform privacy-intruding monitoring, while automated monitoring will be more accurate since computers possess sensory modalities that humans lack" (Pantic and Rothkranz 2001, 466).

18. David's clingy, possessive desire begs the question: "how much do we really want our machines to love us?" At the same time, his growing obsession parallels the conflicted, unrealistic, and contradictory desires of his maker and the world in which he was manufactured. For his creator (Professor Hobbes), David is the embodiment of a well-rehearsed fantasy ("to create an artificial being has been the dream of man since the birth of science"). David's quest to become human represents "part of the great human flaw to wish for things that don't exist, and also the greatest gift—to chase down our dreams." Such attitudes have provided a consistent rationale for roboticists and adventurers alike and yet, in the film, are acted out by one small robot, the "first of a kind" whose journey leads him through decades of industrial waste (comprising mainly old, dysfunctional robots) to derelict theme parks lying at the bottom of the sea—the sea that, thanks to climate change, has since engulfed the city. At the point where David meets his maker, the narrative becomes as excessive as the "dream of science" that inspired the industrial/technological revolution, which led to climate change, which resulted in very sophisticated robots substituting for human children.

19. Sarah Kember's succinct estimation is relevant here: "What succeeded as a short story arguably failed as a film [that was] more preoccupied [with] ... the Oedipal and Pinocchio, than with the current state of Artificial Intelligence" (Kember 2003, 209).

20. Woodward asks the crucial question: "What is the key to believing that a digital life form (a 'bot,' for example) possesses subjectivity?" (Woodward 2004, 191). As Woodward reflects: "The rhetoric of the attribution to and instantiation of emotions in the lifeworld of computers, replicants, and cyborgs, bots and robots, a lifeworld that extends to ours— indeed is ours—serves as ... a coupling device. ... Thus the emotions as thematized in science fiction ... and the emotions as they are experienced in our technological habitat ... serve as a bridge, an intangible but very real prosthesis, one that helps us connect ourselves to the world we have been inventing" (ibid., 192–193).

21. It should be noted here that ambivalence has its critics: on the one hand, it is argued (by Heidegger, Kittler, Virilio, and Derrida) that incorporation is in fact an illusion because we live in a technocracy, a regime where technology shapes and controls all perception and all relationships with media in the interests of capital. On the other, the separation that incorporation implies, even if in the process of dissolution, is critiqued by theorists who situate technics at the origin of the category human and argue that any distinction between the two presupposes a unified body that technology is in the process of enhancing. See for instance John Johnston's critique of Virilio (Johnston 1999, 32).

22. Cited in Lenoir 2002b, 210; emphasis added.

23. "That which bears intelligibility, that which increases intelligibility, is not intelligible—by definition, by virtue of its topological structure. From this standpoint, technics is not intelligible ... it does not belong, by definition, by virtue of its situation, to the field of what it makes possible" (Derrida and Stiegler 2002, 108).

5 Resonance, Anechoica, and Noisy Speech

1. "For it is in actuality that every multiple is haunted by an excess of power that nothing can measure, other than ... a *decision*" (Hallward 2003, 89).

2. There are always more combinations of notes than notes themselves. For this reason, 'music' as a sequence of notes, will always be infinite whereas, tones, as elements of musical harmony, are restricted. The gap between the infinite set [according to Badiou, all human situations are infinite] and its subsets or, as Hallward says, "the excess of parts over elements is properly immeasurable ... [and as such] is the 'real of being-as-being'" (Hallward 2003, 88–89).

3. Brook continues: "And it is finally the throttled *cri* which has the last word, monopolizing the text to the exclusion of the author's own commentary: and underground, half-imprisoned yet infinitely 'free' language, still exercising its unsettling, irregular, malevolent and rebellious magic" (Brook 1992, 19).

4. Also note the Latin *spora*, related to sowing seeds and scattering, and the Greek *sperein*, "to sow seed," and associated with *sporadic* in the French, from *spendein*, "to pour out, make a libation": cf. the French *spondée* (see Partridge 1966, 153, 649).

5. Hallward gives the example of Wagner's *Tristan and Isolde*, "composed at the edge of what is recognizable as 'music': to push any further toward dissonance would have been to leave the classical tonal system altogether" (Hallward 2003, 117).

6. "Names ought to be given according to a natural process, and with a proper instrument, and not at our pleasure" (Plato 1953, 387c).

7. "Sense is consequently not the 'signified' or the 'message': it is *that something like the transmission of a 'message' should be possible*. It is the relation as such and nothing else. Thus, it is as relation that sense configures itself—it configures the *toward* that it is (whereas signification figures itself as identity)" (Nancy 1997, 118).

8. For a critique of Heidegger's use of attunement or *stimmung*, see Dyson 2009a, ch. 2.

9. These include literature, music, sound poetry, radio, theater, dance, film, and new media environments.

10. Excerpt from discussions between Stewart and myself, conducted over the course of 2013 in Sydney, Australia. Unless otherwise noted, all following quotes are from these discussions.

11. In "Vocal Textures," Stewart describes the experience of performing the subject/object works as entering "an immersive field that engenders different mental and physical states or discourses of consciousness. In some senses, I am constantly being deconstructed and reconstructed at the edge of 'composed' subjectivity" (Stewart 2010, 182–183).

12. Stewart continues: "We have the metaphor of parallel processing (yet another computer-based metaphor) but the human voice triggers a multiple or multiply lateral things and it's got the ability to synthesize emotional, psychological, psychoanalytic, musical, and semantic aspects."

13. Stewart continues: "It's incredibly difficult to do, because the breath control is totally different, because there are all these vocal rhythms that are based around speech, which *does* regulate our breathing and pneumonic structures, and making these sounds requires a totally different form of throat, stomach, muscular technique."

14. This would be "an analog voice that is aware of digital possibilities but is embracing them in a different form" (Stewart 2010, 185).

15. As opposed to the way she usually carries herself, "with the subject/object pieces she is much more fluid, and her hands gesticulate as in normal speech."

16. As Stewart comments: "Whenever I came back from Europe I noticed the dismal state of venues in Sydney. The boom in house prices and real estate speculation have driven both artists and venues to the outskirts of the city, or to inner city warehouses under flight paths or in industrial zones."

17. Images available at http://www.mot-art-museum.jp/eng/2012/music/.

18. Nancy (2000, 70) continues: "We are incapable of appropriating this proliferation because we do not know how to think its 'spectacular' nature, which at best gets reduced to a discourse about the uncertain signs of the 'screen' and of 'culture.' ... We are not up to the level of the 'we': we constantly refer ourselves back to a 'sociology' that is itself only the learned form of the 'spectacular market.'"

19. Serres writes: "Before making sense, language makes noise: you can have the latter without the former but not the other way round. After noise, and with the passage of time, a sort of rhythm can develop, an almost recurring movement woven through the fabric of chance. ... In turn, this layer of music, universal before the advent of meaning, carries all meaning within it: distilled, differentiated language selects the meaning or meanings it will isolate from this complex, and then broadcast" (Serres 2008, 120).

20. Healy notes the way that air-conditioning reshaped the culture of the American South by promoting indoor living and eradicating so-called vernacular architecture, such as the use of front porches where people would sit and talk to their neighbors. Not only did the increased use of electricity eventually create conditions that were both environmentally and economically unsustainable, "atmosphere" became conditional: a guarded, normative, controlling, and individualizing possession, a marker of affluence as well as hospitality. The concern with AC-related norms can be traced to a debate (circa 1923) over the character of healthy indoor environments, particularly those of schools, between fresh air and AC advocates. The burgeoning AC industry funded research that, for instance, produced a "comfort chart" defining acceptable temperatures and humidity levels and was successful in replacing the term "fresh air" with "ventilation"—which can be either artificial or natural—in the discussion. The term "fresh air" itself was associated with "support for the open-air crusaders and an anti-technical spirit," removing the common markers of healthy living (fresh air and exercise) from the new matrix of a suitable or "correct" atmosphere based on ratios of volumetric space to amounts of heating and cooling (Healy 2012, 314).

21. "If temporality is the dimension of the subject [ever since St. Augustine, Kant, Husserl, and Heidegger], this is because it defines the subject as what separates *itself*, not only from the other or from the pure 'there,' but also from the self: insofar as it waits for *itself* and retains *itself*, insofar as it desires [itself] and forgets [itself], insofar as it retains, by repeating it, its own empty unity and its projected or ... ejected

unicity. So this sonorous place, space and place—and taking place—as sonority, is not a place where the subject comes to make himself heard [like the concert hall or the studio into which the singer or instrumentalist enters]; on the contrary, it is a place that becomes a subject in so far as sound resounds there" (Nancy 2007, 17).

6 The Racket

1. Most notably, Pierre Henri and Pierre Schaeffer (see Dyson 2009a).

2. The sound of life itself, known through the simple fact of noise made audible—either by silencing the filters that would otherwise mute it or by routing it through transducers—draws Cage dangerously close to the notion of pure experience—in particular, the pure experience of sounds in themselves.

3. Armstrong's questioning of the "virtual community" is relevant here.

"In what ways has the association between networks, telecommunications, and reconfigured concepts of community and connectivity (widely accepted in the literature on 'virtual communities' for example) lent itself to a political figuration or myth, rather than to the necessity of parsing out the grammar—of a simultaneous attachment and detachment, proximity and distance—in which to rethink the ties, webs of relations, and social bonds that articulate an exposure to alterity?" (Armstrong 2009, 87).

4. Hardt (2012) proffers the immaterial as a category in contrast with the distinction between the fictional economy and real economy that is sometimes used to characterize the current phase of capitalism.

5. Note for instance that the term "harmonization" has been adopted by the EU and elsewhere as a euphemism for the neoliberal program of ordering, regulating, and reducing regional differences (which might be considered dissonant) and sovereign control over environmental standards, workers' rights, social services, and so on. According to Susan George, of the top 150 companies in the world, only two are nonfinancial (George 2013).

6. This practice allowed buyers with no documentation and poor credit ratings to be granted mortgages far in excess of their ability to pay, with initial low- or no-interest rate periods disguising the real rate, which would take effect after two or three years.

7. We must not forget that following the dot-com bubble burst, the United States launched its "war on terror," transforming a new economy into a "war economy" where "freedom" stood for democracy itself and, fortunately, or coincidentally, that freedom included the freedom from regulation.

8. As David Harvey and others point out, the current financial crisis is the visible and tangible eruption of neoliberalism, a long-term "project ... designed to restore and consolidate class power ... that coalesced in the 1970s. ... Masked by a lot of

rhetoric about individual freedom, liberty, personal responsibility and the virtues of privatization, the free market and free trade, it legitimized draconian policies designed to restore and consolidate class power" (Harvey 2011, 10).

9. Harvey notes that "within the United States, the geographical constraints on banking were step by step removed from the late 1970s onwards" (Harvey 2011, 19). Deregulation and online trading had the immediate effect of increasing market volatility, as slight fluctuations in the market (market noise) were amplified by automatic trading. In fact, high-frequency trading together with high volatility and a "sell" order issuing from an algorithmic trade have been blamed for the "Flash Crash" of May 2010, when American markets fell by 10 percent in a matter of minutes.

10. Lazzarato continues: "Now the population has only to worry itself with what finance, corporations, and the Welfare State 'externalize' onto society—period!" (Lazzarato 2011, 94).

11. "A wide range of environmental issues, varying from peak oil ... to global warming, have been invoked as underlying explanations for, or at least components of, our current economic difficulties" (Harvey 2011, 73).

12. For Marazzi, the increasing force of the creditor-debtor relationship and integration of monetary, banking, and financial systems under neoliberalism has "made it possible to transform debt into tradable securities on the financial market. [Thus] what is called financialization or 'financialism' represents less of a form of investment financing than an enormous mechanism for managing private and public debt" (Marazzi 2011, 25).

13. Dylan Ratigan, long-time host of CNBC's money show *Fast Money*, has written and spoken extensively on the financial crises since he first became aware of the giant debt bubble in 2007. Ratigan describes the financial crisis, the subsequent bailouts, and the current strategy of printing trillions of dollars, known as "quantitative easing," as the biggest theft and cover-up in American history—a giant Ponzi scheme, where corrupt government, financial institutions, taxpayers, and consumers form something of a hierarchy, with the taxpaying consumer at the base providing capital that flows upward. This tripartite structure is engulfed within an even larger circulation between countries and currencies, regulated less by corrupt government officials than by international diplomatic relations, spheres of influence, global financial entities, and military resources. It would be almost impossible to map the flows of power, money, and influence between the nodes connecting these tripartite systems. However, a necessarily brief and incomplete outline will give an indication of just how complex the processes and mechanisms that produced the financial crisis were, and how, today, these processes are corrupting the administration of the ecology and the economy at a rate and by a magnitude that rising emissions and increases in net wealth are both literally off the charts, and where net wealth is held by a similarly stratospherically diminishing group of individuals—the

"Davos class," as Susan George has named them. Ratigan details the wholesale displacement of a material and substantial bedrock, its reformulation into money (its monetization—bricks and mortar translated into CDOs, or collateralized debt obligations), and its movement beyond the particular state or nation into global financial systems, accompanied by what one could describe as the super-sizing or "weaponizing" of money into political influence and international diplomacy. With deregulation and the creation of CDOs, investment banks intentionally started mixing low-risk and high-risk loans, and, knowing that house prices were overheated and mortgage defaults were going to rise, then insured themselves against losses, so-called derivatives or credit default swaps. With taxpayers guaranteeing their losses, as the government guaranteed the banks in order to allow the banks to loan funds on the basis of future repayment, there was no incentive for the banks to avoid high-risk and potentially high-return investments. Being able to insure themselves against loss and knowing that the bailout would be possible—certainly given their political influence—financial institutions had no limit placed on the amount of risk that they could indulge in. American pension plans were invested in high-risk investments that were falsely rated AAA, and many lost both their homes and their retirement savings. According to Ratigan, both finance and the government benefited: investors and fund managers were attracted by the high returns, governors benefited from the perception that they could expect much higher returns for their state pensions and could therefore increase their budgets, spend more on their constituents, and improve their chances of reelection. Politicians at every level appreciated the political donations, and ordinary citizens found that they could get cheap mortgages and easy credit cards. In 2008, threatening financial Armageddon if they weren't bailed out, the American government adopted the "too big to fail" strategy, which encouraged investment in high-risk/high-return ventures knowing that their losses would be borne by the taxpayer, their profits retained in a classic instance of the neoliberal strategy of socializing the losses and privatizing the gains. Harvey describes what occurred as a "financial coup" that Wall Street "launched against the government and the people of the United States" (Harvey 2011, 5). With the cycle that followed of printing money (or "quantitative easing") to "stabilize" the banking system, the subprime mortgage crisis became a prime mortgage crisis. The "rule of debt"—as something that is owed from the future and rules the present as a form of extraction—became the norm.

14. "Throughout most of history, when overt political conflict between classes did appear, it took the form of pleas for debt cancellation. … What we see in the Bible and other religious traditions are traces of the moral arguments by which such claims were justified, usually subject to all sorts of imaginative twists and turns, but inevitably, to some degree, incorporating the language of the marketplace itself" (Graeber 2012, 87).

15. Or as Lazzarato puts it: "Rights are universal and automatic since they are recognized socially and politically, but debt is administered by evaluating 'morality' and

involves the individual as well as the work on the self which the individual must undertake. The logic of debt now structures and conditions the process of individualization, a constant of social policies" (Lazzarato 2011, 131).

16. "If history shows anything it is that there's no better way to justify relations founded on violence, to make such relations seem moral, than by re-framing them in the language of the debt above all because it's the best way of making the victim seem as if they're doing something wrong" (Graeber 2012, 5).

17. According to Graeber, the core argument of primordial debt theory is that "any attempt to separate monetary policy from social policy is ultimately wrong [since these] have always been the same thing. Governments use taxes to create money and they are able to do so because they have become the guardians of the debt that all citizens have been to one another. This debt is the essence of society itself, it exists long before money and markets, and money and markets themselves are simply ways of chopping pieces of it up. At first, the argument goes, this sense of debt was expressed not through the state but through religion. ... Vedic poems written between 1500 and 1200 BC evince a constant concern with debt which is treated as synonymous with guilt and sin" (Graeber 2012, 56).

18. As Graeber points out, "Exchange implies equality—in dealing with cosmic forces this was simply assumed to be impossible from the start" (Graeber 2012, 63).

19. In the United States, for instance, as more and more homeowners—owing more than their homes were worth—simply "walked away," ignoring warnings of bankruptcy and bad credit ratings and the accusations of "moral hazard" (that, interestingly, began to fade as bank defaults grew), the entire system of debt was shown to be both malleable and ultimately based on law enforcement. From preventing mass foreclosures to negotiating sovereign debt repayments, from the local police to the heads of government, it became obvious that debt is about power. Moving from the economy to the ecology, the same applies: once the facade of individual attempts to "pay off" or "get out of" or "minimize" debt to the environment, once the campaign to individualize ecological degradation is no longer credible (because individual attempts pale to insignificance in the presence of the massive environmental destruction underway in those very sectors of the economy that are perpetuating growth and enabling debt), once the massive power of state and governmental machinery exposes its presence, then the debt to the environment and to future generations ceases to become an individual responsibility. When this occurs, there is no individual "guilt," but rather the battle between sustaining the common and exploiting its every last resource.

20. By casting itself into a generational future—so that debt becomes the responsibility not just of the individual but of generations to come—financialism assumes the theocratic power of governance over time. Since theological time is a present eternity subsumed within the logic of the Infinite (such that all present and future events are knowable, or rather the future is already contained within the present in

the infinite knowledge of God), speculation, futures trading, collateralized debt obligations, risk assessment, and management begin to assume powers once held only by the divine. Whereas life insurance policies have become normalized within the last half-century and, despite their morbid associations, have become almost an accepted part of individual financial management, this new level of what is essentially global "risk management" presents an entirely different face. For generational debt, aligned with generational servitude in the days of slavery and practiced on a global scale, begins to look less like your friendly insurance agent than a massive and unstoppable force, a juggernaut with a name like the IMF, a trinity like Moody's, Standard & Poor's, and Fitch, which operates without reason but within the arcane logic of the markets. There is no access within the daily lives of most people to the decision-making processes of this part-human, part-algorithmic or mathematical, and increasingly powerful conglomerate that operates 24/7, across time zones, and within the microseconds that can influence the outcome of the trade. Like priests of old, we hear an interpretation, we hear in layman's terms the otherwise incomprehensible machinations of financial theory. We begin to realize that it is almost not our place to understand the workings of the global market, the operations of synthetic derivatives, the rises and falls in currencies, the acceptability of QE2 or 3, when, after the Great Depression, the strategy of printing money was roundly critiqued as being inevitably inflationary. With round after round of cuts, we begin to accept the legitimacy of unfairness. The fact that as yet no CEO has been jailed in the United States for his role in the financial collapse while millions of taxpayers and pensioners have seen their taxes rise and pensions obliterated, millions of homeowners have watched their homes taken over by banks which they have bailed out, not just with their own taxes, but with those of their children and their children's children—this grotesque asymmetry, this overriding truth has been accepted almost as fate is accepted, as the whims of the gods of old were accepted, and the necessity for sacrifice built in to the rhythms and traditions of the culture. Infinite time, infinite wealth, infinite power: are these not the characteristics that were once the sole province of the divine? And proceeding with the analogy, is not the "praise" that guarantees not only the power but the intercession of divinity now given by the "experts": the economists, the banks, and most of all the Federal Reserve, whose role it is to intercede on behalf of the populations that they govern?

21. "The series of financial crises has violently revealed a subjective figure that, while already present, now occupies the entirety of public space: the 'indebted man.' The subjective achievements neoliberalism has promised ['everyone a shareholder, everyone an owner, everyone an entrepreneur'] have plunged us into the existential condition of the indebted man, at once responsible and guilty for his particular fate" (Lazzarato 2011, 11).

22. "By training the government to 'promise' [to honor their debt,] capitalism exercises control of the future" (Lazzarato 2011, 46). "The creditor-debtor relation concerns the entirety of the current population as well as the population to come. ...

We are no longer the inheritors of original sin but rather of the debt of preceding generations. … If in times past we were indebted to the community, to the gods, to our ancestors, we are henceforth indebted to the 'God' Capital" (ibid., 32). Lazzarato is quite explicit on this point: "As in theology, where the Holy Trinity encompasses the Father, Son, and Holy Spirit, capital encompasses three different forms—industrial, commercial, and financial" (ibid., 62).

23. Excerpt from a series of interviews conducted by the author with Grace in Sydney, 2012–2013. Unless otherwise noted, all comments by Grace are from these interviews.

24. Grace: "In terms of pitch—there isn't any absolute correlation between the scaling of the emotional indices and the pitch, but there is a numerical correlation based on the division of the octave and the emotional indices."

25. *Articulate Project Space*, Sydney, June 29, 2012.

26. Grace, excerpt from an email to the author, April 2012.

27. Grace: "I would say that all of these developments—say the graphical user interface—aren't artistic developments at all but engineering ones."

28. Some months after the exhibition closed, Grace commented that "none of the works sold. It was an entirely self-financed project, so this work on the self has not been profitable at all—a big loss, really (in terms of the profit-and-loss mentality) but this is the reality for all but a tiny proportion of artists; the art world is the original 1%–99% divide!"

29. We have seen that during the financial crisis, emissions actually went down in the United States for the first time in decades.

Conclusion: Echoes of Eco

1. Audiences, for instance, while noticing the sounds of birds in the background or the rustling of trains that became the "composition" of Cage's *4′33″* in the 1950s, didn't quite know how to react to an environment where aircraft noise obliterated the performance some half a century later. In many ways, the arguments for sound and its inclusion within consciousness and knowledge have taken on the same assumptions of the visual list culture it critiques: perhaps less disembodied, perhaps including the throat and voice but ignoring acoustic space and the space of listening in favor of the particular sonic qualities of the sound produced. Where does the environmental excess or residue go? While focusing on micro-sounds, large sounds, sounds from strange instruments, what happened to the environment that carried them?

2. "Automation House: Confronting Tomorrow's Problems behind Yesterday's Façade," *Interiors* (Nov. 1968). Similarly, the supplement inserted in the *New York Times* on February 1, 1970, demonstrates the importance of this housing:

> Automation House is more than a place. It is a symbol and demonstration of man's wish to shape his future in a world of bewildering change. Individuals everywhere, stimulated through the media of mass communications by sights and sounds not seen nor heard before, yearn for participation and prominence. Instead the new technology creates in man a feeling of isolation and alienation. ... Automation House ... seeks, through the creative use of the tools of technology to give individuals an expanded opportunity for human development. (*Automation House*, New York, American Foundation on Automation and Employment, Advertising supplement inserted the *New York Times* [Feb. 1, 1970])

For more on this topic, see Dyson 2004.

3. As Schwartz mentions, after the Second World War, psychologists and psychophysicists developed an interest in noise that required the production of "machines that did nothing but make noise [defined as sound that was] random and carried no information; it could be heard only as itself." It would be called, however inaccurately, "white noise"—"by inaccurate analogy to the optical white that is all spectra of light, and by improper descent from the 'white noise' of certain singers or passages, defined as pure and colorless tone," Schwartz adds; "it is pure only in the sense that each succeeding sound is innocent of the influence of preceding sounds; it is depleted in the sense that it has no key" (Schwartz 2011, 834).

4. To follow Lazzarato's insistence on calling the series of crises a catastrophe is a point well taken: "From one financial crisis to the next, we have now entered a period of permanent crisis, which we shall call 'catastrophe' to refer to the discontinuity of the concept of crisis itself. Can we still speak of a financial crisis, a nuclear crisis, a food crisis, a climate crisis? Crisis still has a positive connotation. It can refer to a situation capable of being overcome. It has long provided capitalism with the occasion for a new beginning, A New Deal, a new 'pact' for new growth. Today, at least, we have the distinct impression that such is no longer the case, that we have reached a turning point, for the present circumstances look less like a crisis than a catastrophe. ... In modern-day capitalism, 'production' is inseparable from 'destruction.' ... The advances of science simultaneously produce nuclear power capable of destroying several Earth-sized planets; its 'civil' uses pollute the ecosystems beyond human time and force us to live in a permanent state of exception. Industry multiplies the production of consumer goods while at the same time multiplying water, air, and soil pollution and degrading the climate. Agricultural production poisons us at the same time it provides us with food; cognitive capitalism destroys the 'public' education system at every level; cultural capitalism produces historically unprecedented conformism; the image society kills imagination, and so on" (Lazzarato 2011, 151; see also George 2013).

5. "To strip away the sense that has been made in order to allow its sense to come in turn, this is the labor, thought, writing, and extraction that stand before us, their happenstance, their happiness and unhappiness" (Nancy 1997, 150).

6. From http://blogs.aljazeera.net/fault-lines/2011/10/10/ows-human-mic:

> The New York Police Department prohibits the use of electronic sound amplifiers—megaphones, microphones and loudspeakers—without a permit. The occupiers do not have one.
>
> So they are using what they call the human mic.
> It works like this; the person addressing the crowd in the shadow of the large angular sculpture that stands at the corner of Broadway and Cedar and is universally referred to as "the red thing" shouts:
> "It's a beautiful night ..."
> Those seated around her respond by repeating:
> "IT'S A BEAUTIFUL NIGHT ..."
> She goes on:
> "... to occupy Wall Street."
> The echo comes back, much louder, and people who are sitting or standing too far away to hear her solitary voice can hear the words now that they are spoken by hundreds of others:
> "... TO OCCUPY WALL STREET."

According to Graeber, the human mic (which he refers to as the "People's Mic") also nurtures a kind of concise thoughtfulness. Speakers choose their words carefully; rambling is not an option. The assembly listens carefully; you cannot get distracted or talk over the conversation when you have to repeat every word that is spoken. Indeed, the ground rule for the human mic is that everyone must repeat everything that is said, regardless of whether or not you agree with it. In a group of hundreds (or thousands) deprived of megaphones and loudspeakers, it is required in order to hear anything at all, and thus required in order to be able to disagree. So, the human mic seems to cultivate a kind of egalitarian attention to one another. And on occupied Wall Street, what began as a way of circumventing an inconvenient police rule has come to function as a regular demonstration of solidarity and cooperation, amplifying the people's voices (Graeber 2013, 50–51).

7. This was, incidentally, the initial response and persistent criticism of the media coverage of OWS.

8. For instance, Hardt and Negri (2012): "the movements share common characteristics and strategies. They use occupation and encampment, their internal organizer nation is that of a multitude, with horizontal decision-making, no obvious leadership, democratic decision-making and horizontal organizational structures. These characteristics are related to the fundamental driver, the underlying cause, and the material and practical means of organization, which is the Commons."

9. The inward turn operates across registers. With regard to hearing, Nancy writes, "Towards the (pre-philosophical and esoteric Pythagorean) curtain of acousmatics, to the confessional where sin meets forgiveness" (Nancy 2007, 2). With regard to

enclosure, "in terms of the gaze, the subject is referred back to itself as object. In terms of listening, it is, in a way, to itself that the subject refers or refers back." With regard to knowledge, Nancy asks why it is that the ear is involved in "making resonant to, but, in the case of the eye, there is manifestation and display, a making evident?" (ibid.).

10. Reading Nancy, Armstrong cautions that "dialogue" should be translated "not as speech, discourse, or a conversation, that is said to take place between two distinct and separate individuals—a space defined by a sphere of communication regulated by its ideal transparency—but as the rhythmic spacing or interruption of the logos, a rhythmic spacing of all *dia-logue* ... that is now inscribed in and as the sharing out and division of the *ensemble*, of all being-together" (Armstrong 2009, 55).

11. As Armstrong notes, there is "an aporia between a sharing and a simultaneous division of speech, and it is this very aporia that traverses all 'being-together' and 'living-together' at an oblique angle" (Armstrong 2009, 55). This aporia cuts through the assumption of "a" place, a beginning or an interiority that can be owned, enclosed within, and made "a self." It suggests a return to the sound-in-space in which listening occurs—its sonic qualities and the way that these have been interwoven with human activities—but not, however, as a unifying phenomenon. The intimacy of resonance and reverberation, its role in the formation of the subject, in "hearing oneself speak," its associations with understanding, similarity, sympathy, and the social are all *after the fact* of this space where hearing occurs, where sensing takes place. In contradistinction to the materialist orientation of Western metaphysics, there is a kind of material agency in the doubling and reflection of sound and the space it provides for the subject to form. "The sonorous present is the result of space-time: it spreads through space, or rather it opens a space that is its own, the very spreading out of its resonance, its expansion and its reverberation. ... Perhaps it is necessary that sense not be content to make sense (or to be logos), but that it wants also to resound. My whole proposal will revolve around such a fundamental resonance, as a first or last profundity of 'sense' itself (or of truth)" (Nancy 2007, 13, 7).

References

Agamben, Giorgio. 1991. *Language and Death: The Place of Negativity (Theory and History of Literature)*. Trans. Karen E. Pinkus and Michael Hardt. Theory and History of Literature (78). Minneapolis: University of Minnesota Press.

Agamben, Giorgio. 2007. *Infancy and History: On the Destruction of Experience*. London: Verso.

Agamben, Giorgio. 2009. *The Signature of All Things: On Method*. Trans. Luca D'Isanto with Kevin Attell. New York: Zone Books.

Agamben, Giorgio. 2011. *The Kingdom and the Glory: For a Theological Genealogy of Economy and Government (Homo Sacer II, 2)*. Trans. Lorenzo Chiesa. Stanford, CA: Stanford University Press.

Armstrong, Philip. 2009. *Reticulations: Jean-Luc Nancy and the Networks of the Political*. Minneapolis: University of Minnesota Press.

Attali, Jacques. 1985. *Noise: The Political Economy of Music*. Minneapolis: University of Minnesota Press.

Auslander, Philip. 2005. At the *Listening Post*, or Do machines perform? *International Journal of Performance Arts and Digital Media* 1 (1): 6–8.

Automation House. American Foundation on Automation and Employment, New York. Advertising supplement inserted in the *New York Times* (Feb. 1, 1970). http://www.fondation-langlois.org/html/e/page.php?NumPage=2184.

Automation house: Confronting tomorrow's problems behind yesterday's façade. Excerpt from. *Interiors* (Nov. 1968, 115–124). Available at www.fondation-langlois.org/html/e/doc.php?NumEnregDoc+d0009207.

Badiou, Alain. 2007a. *Being and Event*. Trans. Oliver Feltham. London: Continuum.

Badiou, Alain. 2007b. *The Century*. Malden, MA: Polity Press.

Bickmore, T., and J. Cassell. 2005. Social dialogue with embodied conversational agents. In *Natural, Intelligent, and Effective Interaction with Multimodal Dialogue*

Systems, ed. J. van Kuppevelt, L. Dybkjaer, and N. Bernsen. New York: Kluwer Academic.

Bickmore, T., and R. Picard. 2003. Subtle expressivity by relational agents. In *Proceedings of the CHI 2003 Workshop on Subtle Expressivity for Characters and Robots*. April 7, Fort Lauderdale, FL.

Biever, Celeste. 2004. Cellphones turn into smart personal assistants. *New Scientist* 2475 (Nov. 27).

Bouman, Margot. 2003. The machine speaks the world's thoughts. *Parachute* (112): 109–112.

Brook, Peter. 1992. Introduction. In Henri Michaux, *Space, Displaced*, trans. David and Helen Constantine. Northumberland, UK: Bloodaxe Books.

Cage, John. 1961. The future of music: Credo. In *Silence: Lectures and Writings*, ed. John Cage. Middletown, Conn: Wesleyan University Press.

Cahn, Janet. 1988. From sad to glad: Emotional computer voices. In *Proceedings of Speech Tech '88, Voice Input/Output Applications Conference and Exhibition*, 35–37. New York City, April.

Cahn, Janet. 1990. Generation of affect in synthesized speech. *Journal of the American Voice I/O Society* 8:1–19.

Cassell, J. 2000. External manifestations of trustworthiness in the interface. *Association of Computing Machinery* 43 (12): 50–56.

Cassell, J., T. Bickmore, M. Billinghurst, L. Campbell, K. Chang, H. Vilhjálmsson, and H. Yan. 1999. Embodiment in conversational interfaces: Rea. In *Proceedings of the SIGCHI Conference on Human Factors in Computing Systems*. New York: ACM Press.

Cassell, J., T. Bickmore, L. Campbell, H. Vilhjálmsson, and H. Yan. 2001. More than just a pretty face: Conversational protocols and the affordances of embodiment. In *Knowledge-Based Systems* 14 (1): 55–64.

Cavarero, Adriana. 2005. *For More Than One Voice: Toward a Philosophy of Vocal Expression*. Trans. Paul A. Kottman. Stanford: Stanford University Press.

Cohen, Irun R. 2006. Informational landscapes in art, science, and evolution. *Bulletin of Mathematical Biology* 68:1213–1229.

Coyne, Richard. 1999. *Technoromanticism: Digital Narrative, Holism, and the Romance of the Real*. Cambridge, MA: MIT Press.

Deleuze, Gilles. 1993. *The Fold: Leibniz and the Baroque*. Minneapolis: University of Minnesota Press.

Deleuze, Gilles, and Felix Guattari. 1987. *A Thousand Plateaus: Capitalism and Schizophrenia*. Minneapolis: University of Minnesota Press.

Derrida, Jacques, and Bernard Stiegler. 2002. *Echographies of Television: Filmed Interviews*. Malden, MA: Polity Press.

Dolar, Mladen. 2006. *A Voice and Nothing More*. Ed. S. Žižek. Cambridge, MA: MIT Press.

Dreyfus, Hubert. 2001. *On the Internet*. New York: Routledge.

Dyson, Frances. 2004. And then it was now. The Daniel Langlois Foundation for Art, Science, and Technology. Available at http://www.fondation-langlois.org/html/e/page.php?NumPage=2143.

Dyson, Frances. 2009a. *Sounding New Media: Immersion and Embodiment in the Arts and Culture*. Berkeley: University of California Press.

Dyson, Frances. 2009b. Enchanting data: Body, voice, and tone in affective computing. In *Emotion, Place and Culture*, ed. Mick Smith, Joyce Davidson, Laura Cameron, and Liz Bondi, 247–266. Farnham: Ashgate.

Eagle, N., and A. Pentland. 2003a. Social network computing. *Lecture Notes in Computer Science* 2864:289–296.

Eagle, N., and A. Pentland. 2003b. The relationship barometer. Available at http://pubs.media.mit.edu/pubs/papers/TR-567.pdf.

Ekbia, H. R. 2008. *Artificial Dreams: The Quest for Non-Biological Intelligence*. New York: Cambridge University Press.

Gallope, Michael. 2010. The sound of repeating life: Ethics and metaphysics in Deleuze's philosophy of music. In *Sounding the Virtual: Gilles Deleuze and the Theory and Philosophy of Music*, ed. Brian Hulse and Nick Nesbitt, 77–102. Farnham: Ashgate.

George, Susan. 2013. The growing power of illegitimate authority. Wheelwright Memorial Lecture. Co-hosted by the Sydney University Student Union and the Faculty of Arts and Social Sciences, University of Sydney.

Grace, Helen. 2009. IPO: Emotional economies. In *IPO: Helen Grace*, ed. Helen Grace. Exh. cat. Hong Kong: Korkos Gallery; Sydney: Mori Gallery.

Graeber, David. 2012. *Debt: The First 5,000 Years*. Brooklyn: Melville House.

Graeber, David. 2013. *The Democracy Project*. Melbourne, Australia: Allen Lane, Penguin.

Hallward, Peter. 2003. *Badiou: A Subject to Truth*. Foreword by Slavoj Žižek. Minneapolis: University of Minnesota Press.

Hardt, Michael. 2013. Falsify the currency! Talk given at "Unsettled Practices—Work and Expert Knowledge," University of Chicago, Nov. Available at http://www.youtube.com/watch?v=MmZ00vZiJC8.

Hardt, Michael, and Antonio Negri. 2012. *Declaration*. Self-published leaflet, distributed by Argo-Navis.

Harvey, David. 2011. *The Enigma of Capital and the Crises of Capitalism*. New York: Oxford University Press.

Hayles, N. Katherine. 2005. *My Mother Was a Computer: Digital Subjects and Literary Texts*. Chicago: University of Chicago Press.

Healy, Stephen, 2012. Atmospheres of consumption: Shopping and the effects of air-conditioning. In *Emotion, Space, and Society*. Available at http://dx.doi.org/10.1016/j.emospa.2012.10.003.

Heller-Roazen, Daniel. 2011. *The Fifth Hammer: Pythagoras and the Disharmony of the World*. New York: Zone Books.

Hultén, Pontus, and Frank Königsberg. 1966. Catalog entry. In Billy Klüver, *9 Evenings: Theatre and Engineering*, ed. Pontus Hultén and Frank Königsberg, 1. New York: Experiments in Art and Technology; The Foundation for Contemporary Performance Arts.

Johnston, John. 1999. Machinic vision. *Critical Inquiry* 26 (1): 27–48.

Kahn, Douglas. 2004. Ether ore: Mining vibrations for modernist music. In *Hearing Culture: New Directions in the Anthropology of Sound*, ed. Veit Erlmann, 107–130. London: Berg.

Kember, Sarah. 2003. *Cyberfeminism and Artificial Life*. New York: Routledge.

Kepler, Johannes. 2002. *Harmonies of the World* (Book Five). Philadelphia: Running Press.

Klüver, Billy. 1966. *9 Evenings: Theatre and Engineering*. Ed. Pontus Hultén and Frank Königsberg. New York: Experiments in Art and Technology; The Foundation for Contemporary Performance Arts.

Lazzarato, Maurizio. 2011. *The Making of Indebted Man: An Essay on the Neoliberal Condition*. Trans. Joshua David Jordan. Los Angeles: Semiotext(e).

Lenoir, Timothy. 2002a. All but war is simulation: The military-entertainment complex. *Configurations* 8:289–335.

Lenoir, Timothy. 2002b. Makeover: Writing the body into the posthuman technoscape: Part one: Embracing the posthuman. *Configurations* 10:203–220.

Madan, Anmol, Ron Caneel, and Alex Pentland. 2004. Voices of attraction. AugCog Symposium of HCI, MIT Media Lab: Affective Computing Group. Available at http://affect.media.mit.edu/projects.php.

Marazzi, Christian. 2011. *The Violence of Financial Capitalism*. Trans. Kristina Lebedeva and Jason Francis Mc Gimsey. Los Angeles: Semiotext(e).

Michaux, Henri. 1992. *Space, Displaced.* Trans. David Constantine and Helen Constantine. Northumberland, UK: Bloodaxe Books.

Morgan, Margaret. 2009. Face value. In *IPO: Helen Grace,* ed. Helen Grace. Exh. cat. Hong Kong: Korkos Gallery; Sydney: Mori Gallery.

Morton, Timothy. 2012. *The Ecological Thought.* Cambridge, MA: Harvard University Press.

Nancy, Jean-Luc. 1996. The Deleuzian fold of thought. In *Deleuze: A Critical Reader,* ed. Paul Patton. Oxford: Blackwell.

Nancy, Jean-Luc. 1997. *The Sense of the World.* Trans. and with a foreword by Jeffrey S. Librett. Minneapolis: University of Minnesota Press.

Nancy, Jean-Luc. 2000. *Being Singular Plural.* Trans. Robert D. Richardson and Anne E. O'Byrne. Palo Alto: Stanford University Press.

Nancy, Jean-Luc. 2007. *Listening.* Trans. Charlotte Mandell. New York: Fordham University Press.

Nancy, Jean-Luc. 2009. Epigraph to Philip Armstrong, *Reticulations: Jean-Luc Nancy and the Networks of the Political.* Minneapolis: University of Minnesota Press.

Ovid. 1986. *Metamorphosis.* Trans. A. D. Melville with an introduction by E. J. Kenney. Oxford: Oxford University Press.

Pantic, Maja, and Leon Rothkranz. 2001. Affect-sensitive multi-modal monitoring in ubiquitous computing: Advances and challenges. In *ICEIS—Artificial Intelligence and Decision Support Systems,* 466–473.

Partridge, Eric. 1966. *Origins: A Short Etymological Dictionary of Modern English.* New York: Macmillan.

Picard, R. 1997. *Affective Computing.* Cambridge, MA: MIT Press.

Picard, R., and J. Klein. 2002. Computers that recognize and respond to user emotions: Theoretical and practical implications. *Interacting with Computers* 14 (2): 141–169.

Picard, R., E. Vyzas, and J. Healey. 2001. Toward machine emotional intelligence: Analysis of affective physiological state. *IEEE Transactions on Pattern Analysis and Machine Intelligence* 23 (10): 1175–1191.

Plato. 1953. Cratylus. In *The Dialogues of Plato,* vol. 3. Trans. B. Jowett. Oxford: Clarendon Press.

Rubin, Mark, and Ben Hansen. 2002. *Listening Post*: Giving voice to online communication. In *Proceedings of the 2002 International Conference on Auditory Display.* Kyoto, Japan, July 2–5. Available at http://mat.ucsb.edu/~g.legrady/academic/courses/08f200a/hansen_rubin.pdf.

Rubin, Mark, and Ben Hansen. 2003a. *Listening Post.* Available at http://earstudio.com/2010/09/29/listening-post/.

Rubin, Mark, and Ben Hansen. 2003b. *Listening Post,* part 2. Available at http://www.youtube.com/watch?v=QCwfw0v6mlo.

Schafer, R. Murray. 1981. *The Tuning of the World.* Philadelphia: University of Pennsylvania Press.

Scherzinger, Martin. 2010. Enforced deterritorialization, or The trouble with musical politics. In *Sounding the Virtual: Gilles Deleuze and the Theory and Philosophy of Music,* ed. Brian Hulse and Nick Nesbitt. Burlington, VT: Ashgate.

Schwartz, Hillel. 2011. *Making Noise: From Babel to the Big Bang and Beyond.* New York: Zone Books.

Serres, Michel. 1995. *Genesis.* Trans. Genevieve James and James Nielson. Ann Arbor: University of Michigan Press.

Serres, Michel. 2003. *The Natural Contract.* Trans. Elizabeth MacArthur and William Paulson. Ann Arbor: University of Michigan Press.

Serres, Michel. 2008. *The Five Senses: A Philosophy of Mingled Bodies.* Trans. Margaret Sankey and Peter Cowley. London: Continuum.

Serres, Michel. 2011. *Malfeasance: Appropriation through Pollution?* Stanford: Stanford University Press.

Serres, Michael. 2014. *Times of Crisis.* Trans. Anne-Marie Feenberg-Dibon. New York: Bloomsbury.

Stäheli, Urs. 2003. Financial noises: Inclusion and the promise of meaning. In *Soziale Systeme* 9, vol. 2, 244–256. Stuttgart: Lucius and Lucius.

Sterne, Jonathan. 2012. *MP3: The Meaning of a Format.* Durham, NC: Duke University Press.

Stewart, Amanda. 2010. Vocal textures. In *VOICE: Vocal Aesthetics in Digital Arts and Media* , ed. Norie Neumark, Ross Gibson, and Theo van Leeuwen. Cambridge, MA: MIT Press.

Szendy, Peter. 2008. *Listen: A History of Our Ears.* Trans. Charlotte Mandell. New York: Fordham University Press.

Varèse, Edgard. 1967. The liberation of sound. In *Contemporary Composers on Contemporary Music,* ed. Elliott Schwartz and Barney Childs. New York: Holt, Rinehart & Winston.

Vierinen, Juha. 2002. Intelligent software for ubiquitous computing. Helsinki University of Technology, Telecommunications Software and Multimedia Laboratory. Available at http://www.tml.tkk.fi/Studies/Tik-111.590/2001s/papers/juha_vierinen.pdf.

Waldhauer, Fred. 1966. Notes. In *Proportional Control System for the Festival of Art and Engineering. 9 Evenings: Theatre and Engineering fonds.* 9 EV B3-C6. Montreal: The Daniel Langlois Foundation for Art, Science, and Technology.

Woodward, Kathleen. 2004. A "feeling for the cyborg." In *Data Made Flesh: Embodying Information*, ed. Robert Mitchell and Phillip Thurtle. New York: Routledge.

Index